KT-119-306

Springer Undergraduate Mathematics Series

Springer-Verlag London Ltd.

Advisory Board

Other books in this series

Ian Anderson

A First Course in Discrete Mathematics

With 63 Figures

 Springer

Ian Anderson, MA, PhD
Department of Mathematics, University of Glasgow, University Gardens,
Glasgow G12 8QW, UK

Cover illustration elements reproduced by kind permission of:

Aptech Systems, Inc., Publishers of the GAUSS Mathematical and Statistical System, 23804 S.E. Kent-Kangley Road, Maple Valley, WA 98038, USA. Tel: (206) 432 - 7855 Fax (206) 432 - 7832 email: info@aptech.com URL: www.aptech.com

American Statistical Association: Chance Vol 8 No 1, 1995 article by KS and KW Heiner 'Tree Rings of the Northern Shawangunks' page 32 fig 2

Springer-Verlag: Mathematica in Education and Research Vol 4 Issue 3 1995 article by Roman E Maeder, Beatrice Amrhein and Oliver Gloor 'Illustrated Mathematics: Visualization of Mathematical Objects' page 9 fig 11, originally published as a CD ROM 'Illustrated Mathematics' by TELOS: ISBN 0-387-14222-3, German edition by Birkhauser: ISBN 3-7643-5100-4.

Mathematica in Education and Research Vol 4 Issue 3 1995 article by Richard J Gaylord and Kazume Nishidate 'Traffic Engineering with Cellular Automata' page 35 fig 2. Mathematica in Education and Research Vol 5 Issue 2 1996 article by Michael Trott 'The Implicitization of a Trefoil Knot' page 14.

Mathematica in Education and Research Vol 5 Issue 2 1996 article by Lee de Cola 'Coins, Trees, Bars and Bells: Simulation of the Binomial Process' page 19 fig 3. Mathematica in Education and Research Vol 5 Issue 2 1996 article by Richard Gaylord and Kazume Nishidate 'Contagious Spreading' page 33 fig 1. Mathematica in Education and Research Vol 5 Issue 2 1996 article by Joe Buhler and Stan Wagon 'Secrets of the Madelung Constant' page 50 fig 1.

Springer Undergraduate Mathematics Series ISSN 1615-2085

ISBN 978-1-85233-236-5   ISBN 978-0-85729-315-2 (eBook)
DOI 10.107/978-0-85729-315-2

British Library Cataloguing in Publication Data
Anderson, Ian, 1942-
   A first course in discrete mathematics. - (Springer
   undergraduate mathematics series)
   1. Computer science – Mathematics 2. Combinatorial analysis
   I. Title
   510
   ISBN 978-1-85233-236-5

Library of Congress Cataloging-in-Publication Data
Anderson, Ian, 1942-
   A first course in discrete mathematics / Ian Anderson.
      p.   cm. -- (Springer undergraduate mathematics series)
   Includes bibliographical references and index.
   ISBN 978-1-85233-236-5 (alk. paper)
   1. Mathematics. 2. Computer Science – Mathematics. 1. Title. II. Series.
QA39.2.A533   2000
510--dc21   00-063762

Typesetting: Camera ready by the author and Michael Mackey

12/3830-54321 Printed on acid-free paper SPIN 10887527

Preface

This addition to the SUMS series of textbooks is an introduction to various aspects of discrete mathematics. It is intended as a textbook which could be used at undergraduate level, probably in the second year of an English undergraduate mathematics course. Some textbooks on discrete mathematics are written primarily for computing science students, but the present book is intended for students following a mathematics course. The place of discrete mathematics in the undergraduate curriculum is now fairly well established, and it is certain that its place in the curriculum will be maintained in the third millennium.

Discrete mathematics has several aspects. One fundamental part is **enumeration**, the study of counting arrangements of various types. We might count the number of ways of choosing six lottery numbers from $1, 2, \ldots, 49$, or the number of spanning trees in a complete graph, or the number of ways of arranging 16 teams into four groups of four. We develop methods of counting which can deal with such problems.

Next, **graph theory** can be used to model a variety of situations — road systems, chemical molecules, timetables for examinations. We introduce the basic types of graph and give some indication of what the important properties are that a graph might possess.

The third area of discrete mathematics to be discussed in this book is that of **configurations** or **arrangements**. Latin squares are arrangements of symbols in a particular way; such arrangements can be used to construct experimental designs, magic squares and tournament designs. This leads us on to have a look at block designs, which were discussed extensively by statisticians as well as mathematicians due to their usefulness in the design of experiments. The book closes with a brief introduction to the ideas behind error-correcting codes.

The reader does not require a great deal of technical knowledge to be able to cope with the contents of the book. A knowledge of the method of proof by induction, an acquaintance with the elements of matrix theory and of arithmetic modulo n, a familiarity with geometric series and a certain clarity of thought should see the reader through. Often the main problem encountered by the reader is not in the depth of the argument, but in looking at the problem in the "right way". Facility in this comes of course with practice.

Each chapter ends with a good number of examples. Hints and solutions to most of these are given at the end of the book. The examples are a mixture of fairly straightforward applications of the ideas of the chapter and more challenging problems which are of interest in themselves or are of use later on in the book.

My hope is that this text will provide the basis for a first course in discrete mathematics. Obviously the choice of material for such a course is dependent on the interests of the teacher, but there should be enough topics here to enable an appropriate choice to be made. The text has been influenced in countless ways by the many texts that have appeared over the years, but ultimately it is determined by my own preferences, likes and dislikes, and by my own experience of teaching discrete mathematics at different levels over many years, from masterclasses for 14-year-olds to final year honours courses.

I would like to thank the Springer staff for their encouragement to write this book and for their help in its production. Thanks is also due to Gail Henry for converting my manuscript into a LaTeX file, and to Mark Thomson for reading and commenting on many of the chapters.

University of Glasgow, June 2000

Contents

1
Counting and Binomial Coefficients

In this chapter we introduce the basic counting methods, the factorial function and the binomial coefficients. These are of fundamental importance to the subject matter of subsequent chapters. We start with two basic principles.

1.1 Basic Principles

(a) **The multiplication principle.** Suppose that an activity consists of k stages, and that the ith stage can be carried out in α_i different ways, irrespective of how the other stages are carried out. Then the whole activity can be carried out in $\alpha_1 \alpha_2 \dots \alpha_k$ ways.

Example 1.1

A restaurant serves three types of starter, six main courses and five desserts. So a three-course meal can be chosen in $3 \times 6 \times 5 = 90$ ways.

(b) **The addition principle.** If $A_1, \dots, A_k$ are pairwise disjoint sets (i.e. $A_i \cap A_j = \emptyset$ wherever $i \neq j$), then the number of elements in their union is

$$|A_1 \cup \dots \cup A_k| = |A_1| + \dots + |A_k| = \sum_{i=1}^{k} |A_i|.$$

Example 1.2

In the above example, how many different two-course meals (including a main course) are there?

Solution

There are two types of two-course meal to consider. Let A_1 denote the set of meals consisting of a starter and a main course, and let A_2 denote the set of meals consisting of a main course and a dessert. Then the required number is

$$|A_1 \cup A_2| \;=\; |A_1| + |A_2| \qquad \text{(by the addition principle)}$$

$$\;=\; (3 \times 6) + (6 \times 5) \quad \text{(by the multiplication principle)}$$

$$\;=\; 48.$$

1.2 Factorials

How many ways are there of placing a, b and c in a row? There are six ways, namely

$$abc, \; acb, \; bac, \; bca, \; cab, \; cba.$$

Note that there are three choices for the first place, then two for the second, and then just one for the third; so by the multiplication principle there are $3 \times 2 \times 1 = 6$ possible orderings. In general, if we define $n!$ ("n factorial") by

$$n! = n(n-1)(n-2)\ldots 2.1$$

then we have

Theorem 1.1

The number of ways of placing n objects in order is $n!$.

Example 1.3

Four people, A, B, C, D, form a committee. One is to be president, one secretary, one treasurer, and one social convener. In how many ways can the posts be assigned?

Solution

Think of first choosing a president, then a secretary, and so on. There $4! = 24$ possible choices.

The value of $n!$ gets big very quickly:

$$5! = 120, \quad 10! = 3\,628\,800, \quad 50! \cong 3.04 \times 10^{64}.$$

This is an example of what is called *combinatorial explosion*: the number of different arrangements of n objects gets huge as n increases. The enormous size of $n!$ lies behind what is known as the *travelling salesman problem* which will be studied further in Chapter 4. A traveller sets out from home, has to visit n towns and then return home. Given the mileages between the towns, how does the traveller find the shortest possible route? The naive approach of considering each of the $n!$ possible routes is impracticable if n is large, so another approach is needed.

In certain problems, only some of a given set of objects are to be listed.

Example 1.4

A competition on the back of a cereal packet lists ten properties of a car, and asks the consumer to choose the six most important ones, listing them in order of importance. How many different entries are possible?

Solution

There are ten possibilities for the first choice, then nine for the second, and so on down to five for the sixth. So, by the multiplication principle, the number of possible lists is

$$10 \times 9 \times 8 \times 7 \times 6 \times 5 = \frac{10!}{4!} = 151\,200.$$

In general, we have:

Theorem 1.2

The number of ways of selecting r objects from n, in order but with no repetitions, is

$$n(n-1)\dots(n-r+1) = \frac{n!}{(n-r)!}.$$

Example 1.5

A committee of 4 is to be chosen, as in Example 1.3, but this time there are 20 people to choose from. The number of choices of president, secretary, treasurer, social convener is

$$20 \times 19 \times 18 \times 17 = \frac{20!}{16!} = 116\,280.$$

1.3 Selections

Suppose that in Example 1.5 we wanted just the number of ways of choosing four people for a committee, not bothering about the positions they might fill. Denote by $\binom{20}{4}$ (and read as 20-choose-4) the number of ways in which we can choose 4 from 20 where order does not matter.

Each such choice of four from 20 can be ordered in 4! ways to give an assignment to particular positions within the committee, so, by Example 1.5,

$$4! \times \binom{20}{4} = \frac{20!}{16!}.$$

Thus

$$\binom{20}{4} = \frac{20!}{4!16!} = 4845.$$

This argument is general: $\frac{n!}{(n-r)!} = r! \times \binom{n}{r}$, so we have the following general formula.

Theorem 1.3

Let $\binom{n}{r}$ denote the number of unordered selections of r from n where repetitions are not allowed. Then

$$\binom{n}{r} = \frac{n!}{r!(n-r)!}. \tag{1.1}$$

Since some students find the idea of turning ordered selections into unordered selections confusing, here is another way of deriving the formula (1.1).

Suppose we have to choose a team of r players from a pool of n, one of them to be appointed captain. This can be done by first choosing the team - and there are $\binom{n}{r}$ ways of doing this - and then choosing the captain - there are r ways of doing this. So there are $r\binom{n}{r}$ choices altogether. But we could, instead, first choose the captain - there are n ways of doing this - and then choose the $r - 1$ other members of the team - and there are $\binom{n-1}{r-1}$ ways of doing this. So the number of choices is also $n\binom{n-1}{r-1}$. Thus

$$r\binom{n}{r} = n\binom{n-1}{r-1},$$

so that

$$\binom{n}{r} = \frac{n}{r}\binom{n-1}{r-1}. \tag{1.2}$$

But similarly, $\binom{n-1}{r-1} = \frac{n-1}{r-1}\binom{n-2}{r-2}$, on replacing n by $n-1$ and r by $r-1$, so we get

$$\binom{n}{r} = \frac{n}{r} \cdot \frac{n-1}{r-1}\binom{n-2}{r-2}.$$

Continuing in this way we obtain

$$\binom{n}{r} = \frac{n}{r} \cdot \frac{n-1}{r-1} \cdots \frac{n-(r-2)}{2} \left(\frac{n-(r-1)}{1}\right).$$

Since $\binom{m}{1}$ is clearly always m, we obtain finally

$$\binom{n}{r} = \frac{n(n-1)\ldots(n-r+1)}{r!} = \frac{n!}{r!(n-r)!}$$

as before.

[Note in passing that this is a good example of the technique of **counting the same thing in two different ways**.]

Example 1.6

In the UK National Lottery, a participant chooses six of the numbers 1 to 49; order does not matter. So the number of possible choices is

$$\binom{49}{6} = \frac{49 \times 48 \times 47 \times 46 \times 45 \times 44}{6!} = 13\,983\,816.$$

So there is roughly one chance in 14 million of winning the jackpot!

Example 1.7

How likely is it that next week's lottery winning numbers will be disjoint from this week's?

Solution

There are $\binom{49}{6}$ possible selections next week. The number of these which are disjoint from this week's must be $\binom{43}{6}$, since six of the 49 numbers are ruled out. Since all $\binom{49}{6}$ selections are equally likely, the required probability is

$$\binom{43}{6} \Big/ \binom{49}{6} = 0.436\ldots.$$

Example 1.8

Binary sequences. There are 2^n n-digit binary sequences since each of the n digits is 0 or 1. For example, the eight binary sequences of length three are

$$000 \quad 001 \quad 010 \quad 011 \quad 100 \quad 101 \quad 110 \quad 111.$$

(a) How many binary sequences of length 12 contain exactly six 0s?

(b) How many have more 0s than 1s?

Solution

(a) The six 0s occupy six of the 12 positions. There are $\binom{12}{6} = 924$ choices of these six positions, and this gives the number required.

(b) There are $2^{12} - 924 = 3172$ sequences with unequal numbers of 0s and 1s. By symmetry, exactly half of these, i.e. 1586, will have more 0s than 1s.

We close this section with two simple properties of the numbers $\binom{n}{r}$ which arise out of the fact that $\binom{n}{r}$ is the number of ways of choosing r from n. From now on, we follow the conventions that $0! = 1$ and that $\binom{n}{0} = 1$ for all $n \geq 0$.

Theorem 1.4

(i) $\binom{n}{r} = \binom{n}{n-r}$ for all r, $0 \leq r \leq n$.

(ii) $\binom{n+1}{r} = \binom{n}{r} + \binom{n}{r-1}$ for all r, $0 < r \leq n$.

Proof

(i) This follows immediately from the formula (1.1); alternatively, simply observe that choosing r from n is just the same as selecting the $n - r$ which are not to be chosen!

(ii) A choice of r of the $n + 1$ objects $x_1, \ldots, x_{n+1}$ may or may not include x_{n+1}. If it does not, then r objects have to be chosen from $x_1, \ldots, x_n$ and there are $\binom{n}{r}$ such choices. If it does contain x_{n+1}, then $r - 1$ further objects have to be chosen from $x_1, \ldots, x_n$, and there are $\binom{n}{r-1}$ such choices. The result now follows from the addition principle.

Alternatively we can use the formula (1.1):

$$\binom{n}{r} + \binom{n}{r-1} = \frac{n!}{r!(n-r)!} + \frac{n!}{(r-1)!(n-r+1)!}$$

$$= \frac{n!(n-r+1)}{r!(n-r+1)!} + \frac{n!\,r}{r!(n-r+1)!}$$

$$= \frac{n!}{r!(n-r+1)!}\{n-r+1+r\} = \frac{(n+1)!}{r!(n+1-r)!} = \binom{n+1}{r}.$$

1.4 Binomial Coefficients and Pascal's Triangle

The choice numbers $\binom{n}{r}$ are known as **binomial coefficients**. In this section we find out why. Note that

$$
\begin{aligned}
(1+y)^0 &= 1 \\
(1+y)^1 &= 1+y \\
(1+y)^2 &= 1+2y+y^2 \\
(1+y)^3 &= 1+3y+3y^2+y^3 \\
(1+y)^4 &= 1+4y+6y^2+4y^3+y^4 \\
&\vdots
\end{aligned}
$$

coefficients:

$$
\begin{array}{ccccccccc}
 & & & & 1 & & & & \\
 & & & 1 & & 1 & & & \\
 & & 1 & & 2 & & 1 & & \\
 & 1 & & 3 & & 3 & & 1 & \\
1 & & 4 & & 6 & & 4 & & 1
\end{array}
$$

Note the triangular array of coefficients, known as **Pascal's triangle**. Each row consists of the choice numbers: e.g. the bottom row shown consists of

$$\binom{4}{0} = 1, \quad \binom{4}{1} = 4, \quad \binom{4}{2} = 6, \quad \binom{4}{3} = 4, \quad \binom{4}{4} = 1.$$

Pascal (1623–1662) certainly studied these numbers, making use of them in his work on probability, but the triangle was known to Chinese mathematicians long before.

Theorem 1.5

The binomial theorem

$$(x+y)^n = \binom{n}{0}x^n + \binom{n}{1}x^{n-1}y + \binom{n}{2}x^{n-2}y^2 + \cdots + \binom{n}{n}y^n = \sum_{r=0}^{n}\binom{n}{r}x^{n-r}y^r.$$

Proof

$(x + y)^n = (x + y)(x + y) \cdots (x + y)$ (n brackets). So the coefficient of $x^{n-r}y^r$ in the expansion is the number of ways of getting $x^{n-r}y^r$ when the n brackets are multiplied out. Each term in the expansion is the product of one term from each bracket; so $x^{n-r}y^r$ is obtained as many times as we can choose y from r of the brackets (and x from the remaining $n - r$ brackets). But this is just the number of ways of choosing r of the n brackets, which is $\binom{n}{r}$.

Corollary 1.6

$(1 + y)^n = \sum_{r=0}^{n}\binom{n}{r}y^r$.

Example 1.9

$1 + \binom{10}{1}2 + \binom{10}{2}2^2 + \cdots + \binom{10}{10}2^{10} = (1 + 2)^{10} = 3^{10}$.

Pascal's triangle

```
                    1                          ← row 0
                 1     1                        ← row 1
              1     2     1
           1     3     3     1
        1     4     6     4     1
     1     5    10    10     5     1
  1     6    15    20    15     6     1
1     7    21    35    35    21     7     1      ← row 7
```

In the nth row, $n \geq 0$, the entries are the binomial coefficients $\binom{n}{r}, 0 \leq r \leq n$. The triangle displays the two properties of Theorem 1.4: (i) is shown in the reverse symmetry of each row, and (ii) is shown by the fact that each entry in the triangle is the sum of the two entries immediately above it (e.g. $21 = 6 + 15$). Note also the entries in each row add up to a power of 2.

Theorem 1.7

(i) $\dbinom{n}{0} + \dbinom{n}{1} + \dbinom{n}{2} + \cdots + \dbinom{n}{n} = 2^n;$

(ii) $\dbinom{n}{0} - \dbinom{n}{1} + \dbinom{n}{2} - \cdots + (-1)^n \dbinom{n}{n} = 0 \quad (n > 0).$

Proof

(i) Put $y = 1$ in Corollary 1.6. (ii) Put $y = -1$.

The next result establishes a pattern relating to the **diagonals** in Pascal's triangle. Note, for example, that in a line parallel to the left side of the triangle, we have $1 + 3 + 6 + 10 + 15 = 35$.

Theorem 1.8

For all $m \geq 0$ and $n \geq 1$,

$$\binom{m}{m} + \binom{m+1}{m} + \cdots + \binom{m+n}{m} = \binom{m+n+1}{m+1}.$$

Proof

$$
\begin{aligned}
\binom{m+n+1}{m+1} &= \binom{m+n}{m} + \binom{m+n}{m+1} \quad \text{by Theorem 1.4(ii)} \\[2mm]
&= \binom{m+n}{m} + \binom{m+n-1}{m} + \binom{m+n-1}{m+1} \quad \text{again by 1.4(ii)} \\[2mm]
&= \binom{m+n}{m} + \binom{m+n-1}{m} + \cdots + \binom{m+1}{m} + \binom{m+1}{m+1} \\[2mm]
&= \binom{m+n}{m} + \binom{m+n-1}{m} + \cdots + \binom{m+1}{m} + \binom{m}{m},
\end{aligned}
$$

since $\binom{m+1}{m+1} = \binom{m}{m} = 1$.

Identities

The binomial theorem can be used to obtain other identities involving binomial coefficients.

Example 1.10

Consider the identity

$$(1+x)^n(1+x)^n = (1+x)^{2n}$$

i.e.

$$\left\{\binom{n}{0} + \binom{n}{1}x + \cdots + \binom{n}{n}x^n\right\}\left\{\binom{n}{0} + \binom{n}{1}x + \cdots + \binom{n}{n}x^n\right\} = \sum_{r=0}^{2n}\binom{2n}{r}x^r.$$

Equating the coefficients of x^n on each side of this identity gives

$$\binom{n}{0}\binom{n}{n} + \binom{n}{1}\binom{n}{n-1} + \cdots + \binom{n}{n}\binom{n}{0} = \binom{2n}{n}$$

which, by Theorem 1.4(i) can be rewritten as

$$\binom{n}{0}^2 + \binom{n}{1}^2 + \cdots + \binom{n}{n}^2 = \binom{2n}{n}.$$

For example, with $n = 4$ we get

$$1^2 + 4^2 + 6^2 + 4^2 + 1^2 = 70 = \binom{8}{4}.$$

A similar type of argument enables us to obtain a corresponding result for alternating sums.

Example 1.11

Use the identity $(1 - x^2)^n = (1 - x)^n(1 + x)^n$ to sum

$$\binom{n}{0}^2 - \binom{n}{1}^2 + \binom{n}{2}^2 - \cdots.$$

Solution

Consider the coefficient of x^n on both sides of the given identity.
   Since the right-hand side is

$$\left(1 - \binom{n}{1}x + \binom{n}{2}x^2 - \cdots + (-1)^n\binom{n}{n}x^n\right)\left(1 + \binom{n}{1}x + \cdots + \binom{n}{n}x^n\right),$$

the coefficient of x^n is

$$\sum_{r+s=n}(-1)^r\binom{n}{r}\binom{n}{s} = \sum_{r=0}^{n}(-1)^r\binom{n}{r}\binom{n}{n-r} = \sum_{r=0}^{n}(-1)^r\binom{n}{r}^2.$$

The coefficient of x^n on the left of the given identity is 0 if n is odd (why?) but is $(-1)^{\frac{n}{2}} \begin{pmatrix} n \\ \frac{n}{2} \end{pmatrix}$ if n is even. So we have

$$\binom{n}{0}^2 - \binom{n}{1}^2 + \binom{n}{2}^2 - \cdots + (-1)^n \binom{n}{n}^2 = \begin{cases} (-1)^{\frac{n}{2}} \binom{n}{\frac{n}{2}} & n \text{ even} \\ 0 & n \text{ odd.} \end{cases}$$

To illustrate:

$$\binom{5}{0}^2 - \binom{5}{1}^2 + \binom{5}{2}^2 - \binom{5}{3}^2 + \binom{5}{4}^2 - \binom{5}{5}^2 = 0 \quad \text{(obviously!)};$$

$$\binom{4}{0}^2 - \binom{4}{1}^2 + \binom{4}{2}^2 - \binom{4}{3}^2 + \binom{4}{4}^2 = 1 - 16 + 36 - 16 + 1 = 6 = \binom{4}{2};$$

$$\binom{6}{0}^2 - \binom{6}{1}^2 + \binom{6}{2}^2 - \binom{6}{3}^2 + \binom{6}{4}^2 - \binom{6}{5}^2 + \binom{6}{6}^2 = -20 = -\binom{6}{3}.$$

1.5 Selections with Repetitions

We have already seen that there are 2^m binary sequences of length m. Here we are choosing m digits in order, and there are two choices (0 or 1) for each.

Example 1.12

The number of subsets of a set of m elements is 2^m. For each subset corresponds to a binary sequence of length m, in which the ith digit is 1 precisely when the ith element is in the subset. For example, the subset $\{2, 3, 5\}$ of $\{1, 2, 3, 4, 5\}$ corresponds to 01101. Which subsets are represented by (a) 11100, (b) 00000? $[\{1, 2, 3\}, \emptyset]$.

Example 1.13

Each weekend in February I can visit any one of three cinemas. How many different sequences of visits are possible, repeat visits of course being allowed?

Solution

Each weekend I have 3 choices, so by the multiplication principle the total number of visiting sequences is $3^4 = 81$.

Clearly we have

Theorem 1.9

The number of ways of choosing r objects from n, in order and with repetitions allowed, is n^r.

Suppose now that we have to choose r objects from n, where repetitions are allowed, but where order does not matter.

Example 1.14

There are ten ways of choosing two objects from $\{1, 2, 3, 4\}$ unordered, with repetitions allowed. They are:

$$1,1 \quad 1,2 \quad 1,3 \quad 1,4 \quad 2,2 \quad 2,3 \quad 2,4 \quad 3,3 \quad 3,4 \quad 4,4.$$

Theorem 1.10

The number of unordered choices of r from n, with repetitions allowed is

$$\binom{n+r-1}{r}.$$

Proof

Any choice will consist of x_1 choices of the first object, x_2 choices of the second object, and so on, subject to the condition $x_1 + \cdots + x_n = r$. So the required number is just the number of solutions of the equation $x_1 + \cdots + x_n = r$ in non-negative integers x_i.

Now we can represent a solution $x_1, \ldots, x_n$ by a binary sequence:

$$x_1 0\text{s}, 1, x_2 0\text{s}, 1, x_3 0\text{s}, 1, \ldots, 1, x_n 0\text{s}.$$

Think of the 1s as indicating a move from one object to the next. For example, the solution $x_1 = 2, x_2 = 0, x_3 = 2, x_4 = 1$ of $x_1 + x_2 + x_3 + x_4 = 5$ corresponds to the binary sequence 00110010. Corresponding to $x_1 + \cdots + x_n = r$, there will be $n - 1$ 1s and r 0s, and so each sequence will be of length $n + r - 1$, containing exactly r 0s. Conversely, any such sequence corresponds to a non-negative integer solution of $x_1 + \cdots + x_n = r$. Now the r 0s can be in any of the $n + r - 1$ places, so the number of such sequences, and hence the number of unordered choices, is $\binom{n+r-1}{r}$, the number of ways of choosing r places out of $n + r - 1$.

This proof also establishes the following result.

Theorem 1.11

The number of solutions of $x_1 + \cdots + x_n = r$ in non-negative integers x_i is

$$\binom{n+r-1}{r}.$$

Example 1.15

The number of solutions of $x + y + z = 17$ in non-negative integers is

$$\binom{17 + 3 - 1}{17} = \binom{19}{17} = \binom{19}{2} = 171.$$

Example 1.16

How many solutions are there of $x + y + z = 17$ in **positive** integers?

Solution

Here we require $x \geq 1, y \geq 1, z \geq 1$, so we put $x = 1 + u, y = 1 + v, z = 1 + w$. The equation becomes $u + v + w = 14$, and we seek solutions in non-negative integers u, v, w. The number of solutions is therefore

$$\binom{14 + 3 - 1}{14} = \binom{16}{2} = 120.$$

Example 1.17

How many binary sequences are there containing exactly p 0s and q 1s, $q \geq p - 1$ with no two 0s together?

Solution

Imagine the q 1s placed in a row: $\ldots 1 \ldots 1 \ldots 1 \ldots$. They create $q + 1$ "boxes" (spaces) in which to put the 0s ($q - 1$ boxes between the 1s, and one at each end). The p 0s have to be places in **different** boxes, so the number of choices is $\binom{q+1}{p}$.

Alternatively, we could first place the p 0s in a row. They create $p + 1$ boxes in which to place the q 1s. But the $p - 1$ internal boxes must receive at least one 1. If we let x_i denote the number of 1s placed in the ith box, we want the number of solutions of $x_1 + \cdots + x_{p+1} = q$, where $x_1 \geq 0$, $x_{p+1} \geq 0$, and all other $x_i \geq 1$. Putting $x_1 = y_1$, $x_{p+1} = y_{p+1}$, $x_i = 1 + y_i$ otherwise, the equation becomes $y_1 + \cdots + y_{p+1} = q - (p - 1) = q - p + 1$, and the number of non-negative integer solutions is

$$\binom{q - p + 1 + p + 1 - 1}{q - p + 1} = \binom{q + 1}{q + 1 - p} = \binom{q + 1}{p}.$$

We summarise the selection formulae in Table 1.1.

Table 1.1 Summary of formulae for choosing r from n.

choose r from n	ordered	unordered
no repetitions allowed	$\dfrac{n!}{(n-r)!}$	$\dbinom{n}{r}$
repetitions allowed	n^r	$\dbinom{n+r-1}{r}$

1.6 A Useful Matrix Inversion

In this final section we present an elegant and useful matrix result which will be of use later. It enables us to deduce, from a relation of the form $a_n = \sum_k \binom{n}{k} b_k$, an expression for b_n in terms of the a_i.

To begin, first consider the matrix

$$\begin{bmatrix} 1 & 0 & 0 & 0 \\ 1 & 1 & 0 & 0 \\ 1 & 2 & 1 & 0 \\ 1 & 3 & 3 & 1 \end{bmatrix}$$

which is clearly constructed from the first four rows of Pascal's triangle. Remarkably, its inverse is

$$\begin{bmatrix} 1 & 0 & 0 & 0 \\ -1 & 1 & 0 & 0 \\ 1 & -2 & 1 & 0 \\ -1 & 3 & -3 & 1 \end{bmatrix}$$

since the product of these two matrices is easily checked to be the identity. This result generalises in the obvious way. Before proving this, we need a couple of lemmas.

Lemma 1.12

For all i, j, k, $j \le k \le i$,

$$\binom{i}{k}\binom{k}{j} = \binom{i}{j}\binom{i-j}{k-j}.$$

Proof

$$\binom{i}{k}\binom{k}{j} = \frac{i!}{k!(i-k)!}\frac{k!}{j!(k-j)!} = \frac{i!}{(i-k)!j!(k-j)!}$$

$$= \frac{i!}{j!(i-j)!}\frac{(i-j)!}{(i-k)!(k-j)!}$$

$$= \binom{i}{j}\binom{i-j}{k-j}.$$

Lemma 1.13

$\sum_{j\leq k\leq i}\binom{i}{k}\binom{k}{j}(-1)^k = 0$ whenever $j < i$.

Proof

$$\sum_{j\leq k\leq i}\binom{i}{k}\binom{k}{j}(-1)^k = \sum_{j\leq k\leq i}\binom{i}{j}\binom{i-j}{k-j}(-1)^k \text{ by Lemma 1.12}$$

$$= (-1)^j\binom{i}{j}\sum_{0\leq k-j\leq i-j}\binom{i-j}{k-j}(-1)^{k-j}$$

$$= (-1)^j\binom{i}{j}\sum_{\ell=0}^{i-j}\binom{i-j}{\ell}(-1)^{\ell} \quad \text{on putting } \ell = k-j$$

$$= (-1)^j\binom{i}{j}(1-1)^{i-j} = 0 \quad \text{since } j < i.$$

We extend the definition of $\binom{i}{j}$ to cases where $i < j$ by putting $\binom{i}{j} = 0$. This makes sense, since there is no way of choosing j objects from $i < j$ without repetitions.

Theorem 1.14

Let A be the $(n+1) \times (n+1)$ matrix, with rows and columns labelled by $0, 1, \ldots n$, defined by $a_{ij} = \binom{i}{j}$. Let B be the $(n+1) \times (n+1)$ matrix defined by $b_{ij} = (-1)^{i+j}\binom{i}{j}$. Then $BA = I$.

Proof

The (i,i)th entry of BA is the product of the ith row of B and the ith column of A, and so is $\sum_k b_{ik}a_{ki} = \sum_k (-1)^{i+k}\binom{i}{k}\binom{k}{i}$. There is only one value of k for which $\binom{i}{k}$ and $\binom{k}{i}$ are both nonzero, namely $k = i$, so the sum reduces to $(-1)^{2i}\binom{i}{i}\binom{i}{i} = 1$. So all diagonal entries of BA are 1.

If $i \neq j$, the (i,j)th entry of BA is $\sum_k (-1)^{i+k} \binom{i}{k} \binom{k}{j} = (-1)^i \sum_k \binom{i}{k} \binom{k}{j} (-1)^k$. This is 0, by Lemma 1.12, whenever $i > j$. But if $i < j$ then every term $\binom{i}{k} \binom{k}{j}$ is 0, so again the sum is 0.

So, now that Theorem 1.14 is established, suppose we are given two sequences $(a_0, a_1, a_2, \dots)$ and $(b_0, b_1, b_2, \dots)$ related by

$$a_n = \sum_{k=0}^{n} \binom{n}{k} b_k$$

for all $n \geq 0$. We then have the matrix identity

$$\begin{bmatrix} a_0 \\ a_1 \\ \vdots \\ a_n \end{bmatrix} = A \begin{bmatrix} b_0 \\ b_1 \\ \vdots \\ b_n \end{bmatrix}$$

so by Theorem 1.14 we can deduce that

$$\begin{bmatrix} b_0 \\ b_1 \\ \vdots \\ b_n \end{bmatrix} = BA \begin{bmatrix} b_0 \\ b_1 \\ \vdots \\ b_n \end{bmatrix} = B \begin{bmatrix} a_0 \\ a_1 \\ \vdots \\ a_n \end{bmatrix}.$$

The bottom row then gives

$$b_n = \sum_{k=0}^{n} (-1)^{n+k} \binom{n}{k} a_k.$$

We therefore have the following useful inversion.

Corollary 1.15

If $a_n = \sum_{k=0}^{n} \binom{n}{k} b_k$ for all $n \geq 0$, then $b_n = \sum_{k=0}^{n} (-1)^{n+k} \binom{n}{k} a_k$.

We shall use this inversion result in Sections 2.4 and 5.3.

Exercises

Exercise 1.1

The 10 cabinet ministers of Newland sit around a circular table. One seat is reserved for the Prime Minister. In how many ways can the remaining nine seat themselves?

Exercise 1.2

An 8-person committee is to be formed from a group of 15 women and 12 men. In how many ways can the committee be chosen if
(a) the committee must contain four men and four women?
(b) there must be at least two men?
(c) there must be more women than men?

Exercise 1.3

An eccentric, Mr. Leuben, is reported to have bet that if he shuffled a pack of cards long enough the cards would eventually appear in a given order. He tried for 10 hours a day for 20 years before eventually succeeding after 4 146 028 shuffles. Was he lucky?

Exercise 1.4

Find the probabilities of getting exactly (a) 3, (b) 4, (c) 5 numbers correct in the UK National Lottery.

Exercise 1.5

Estimate your chances of picking the winning numbers in the following lotteries:

(a) Sweden - choose 7 from 35; (b) Hungary - choose 5 from 90.

Exercise 1.6

In the Thunderball variation of the lottery, you choose five from 1 to 34, and one from 1 to 14. Compare your chance of winning Thunderball with your chances of winning the UK Lottery.

Exercise 1.7

What is the probability of getting exactly 5 heads in 10 tosses of a coin?

Exercise 1.8

Give an induction proof of the binomial theorem. (You will need to use Theorem 1.4(ii).)

Exercise 1.9

Deduce from Theorem 1.7 that $\binom{n}{0} + \binom{n}{2} + \binom{n}{4} + \cdots = 2^{n-1}$.

Exercise 1.10

Use the identity $(1+x)^r(1+x)^s = (1+x)^{r+s}$ to derive the **Vandermonde identity**, $\sum_k \binom{r}{k}\binom{s}{n-k} = \binom{r+s}{n}$.

Deduce that $\sum_k \binom{r}{k}\binom{s}{k-1} = \binom{r+s}{r-1}$.

Exercise 1.11

(a) Check that $\binom{7}{0} - \binom{7}{1} + \binom{7}{2} - \binom{7}{3} = -20 = -\binom{6}{3}$.

(b) Generalise this result to $\binom{n}{0} - \binom{n}{1} + \binom{n}{2} - \cdots + (-1)^k\binom{n}{k} = (-1)^k\binom{n-1}{k}$ by replacing each $\binom{n}{i}$ by $\binom{n-1}{i} + \binom{n-1}{i-1}$.

(c) Give an alternative proof by considering the coefficient of x^k on both sides of the identity $(1+x)^{n-1} = (1+x)^n(1+x)^{-1}$.

Exercise 1.12

Use (1.2) to prove that $\sum_k k\binom{m}{k}\binom{n}{k} = n\binom{m+n-1}{n}$.

Exercise 1.13

Find the number of solutions of the equation $x + y + z + w = 15$ (a) in non-negative integers, (b) in positive integers, (c) in integers satisfying $x > 2, y > -2, z > 0, w > -3$.

Exercise 1.14

Find the number of non-negative integer solutions of $x_1 + x_2 + x_3 + x_4 \leq 6$.

Exercise 1.15

Show that more than half of the selections of 6 from 49 in the UK Lottery have no two consecutive numbers.

Exercise 1.16

Find the number of ways of placing 4 marbles in 10 distinguishable boxes if

(a) the marbles are distinguishable, and no box can hold more than one marble;

(b) the marbles are indistinguishable, and no box can hold more than one;

(c) the marbles are distinguishable, and each box can hold any number of them;

(d) the marbles are indistinguishable, and each box can hold any number of them.

Exercise 1.17

Using (1.2), prove that $\sum_{r=0}^{n} r\binom{n}{r} = n2^{n-1}$. Why is this result obvious when we consider the average size of a subset of a set of n elements?

Exercise 1.18

Establish the formula $1^2 + 2^2 + \cdots + n^2 = \frac{1}{6}n(n+1)(2n+1)$ by the following methods.

(a) Prove that $\binom{r+2}{3} - \binom{r}{3} = r^2$, and sum over r.

(b) Prove that $\binom{r}{2} + \binom{r+1}{2} = r^2$, and sum over r, using Theorem 1.8.

Exercise 1.19

(For those who know de Moivre's theorem.)

Let $w = \frac{1}{2}(-1+i\sqrt{3})$, so that $w^3 = 1$ and $w^2+w+1 = 0$. Put $x = 1, y = w$ in the binomial theorem to show that

$$\binom{n}{0} + \binom{n}{3} + \binom{n}{6} + \cdots = \frac{1}{3}\left(2^n + 2\cos\frac{n\pi}{3}\right).$$

(Hint: $1 + w = e^{\frac{\pi i}{3}}$.)

2
Recurrence

It often happens that, in studying a sequence of numbers a_n, a connection between a_n and a_{n-1}, or between a_n and several of the previous $a_i, i < n$, is obtained. This connection is called a recurrence relation; it is the aim of this chapter to illustrate how such recurrences arise and how they may be solved.

2.1 Some Examples

Example 2.1 (The towers of Hanoi)

We begin with a problem made famous by the nineteenth century French mathematician E. Lucas. Consider n discs, all of different sizes, with holes at their centres (like old gramophone records), and three vertical poles onto which the discs can be slipped. Initially all the discs are on one of the poles, in order of size, with the largest at the bottom, forming a tower. It is required to move the discs, one at a time, finishing up with the n discs similarly arranged on one of the other poles. There is the important requirement that at no stage may any disc be placed on top of a smaller disc. What is the minimum number of moves required?

Let a_n denote the smallest number of moves required to move the n discs. Then clearly $a_1 = 1$. Also, $a_2 = 3$: move the top disc to one pole, the bottom to the other, and then place the smaller on top of the larger. What about a_n? It should be clear that, to be able to move the bottom disc, there has to be an empty pole to move it to, and so all the other $n-1$ discs must have been moved to the third pole. To get to this stage, a_{n-1} moves are needed. The largest disc is then moved to the free pole, and then another a_{n-1} moves can position the other $n-1$ discs on top of it. So

$$a_n = 2a_{n-1} + 1.$$

This **recurrence relation**, along with the **initial condition** $a_1 = 1$, enables us to find a_n. We have $a_2 = 2.1 + 1 = 3, a_3 = 2.3 + 1 = 7, a_4 = 2.7 + 1 = 15$, and it appears that $a_n = 2^n - 1$. This can be confirmed by induction, or by iteration:

$$
\begin{aligned}
a_n &= 1 + 2a_{n-1} = 1 + 2(1 + 2a_{n-2}) = 1 + 2 + 2^2 a_{n-2} \\
&= 1 + 2 + 2^2(1 + 2a_{n-3}) = 1 + 2 + 2^2 + 2^3 a_{n-3} \\
&= 1 + 2 + 2^2 + \cdots + 2^{n-2} + 2^{n-1} a_1 \\
&= 1 + 2 + 2^2 + \cdots + 2^{n-1} = 2^n - 1.
\end{aligned}
$$

In the mythical story attached to the puzzle, n was 64 and priests had to move discs of pure gold; when all was accomplished, the end of the world would come. But $2^{64} - 1 = 18\,446\,744\,073\,709\,551\,615$, and at one move per second the process would take about 5.82×10^{11} years; so we have nothing to worry about! This is another good example of combinatorial explosion.

Example 2.2

There are 3^n n-digit sequences in which each digit is $0, 1$ or 2. How many of these sequences have an **odd** number of 0s?

Solution

Let b_n denote the number of such sequences of length n with an odd number of 0s. Each such sequence ends in $0, 1$ or 2. A sequence ending in 1 has any of the b_{n-1} sequences of length $n-1$ preceding the 1; and similarly there are b_{n-1} sequences ending in 2. If a sequence ends in 0, the 0 must be preceded by a sequence of length $n-1$ with an **even** number of 0s; but the number of such sequences is 3^{n-1} (the total number of sequences of length $n-1$) minus b_{n-1} (the number of sequences of length $n-1$ with an **odd** number of 0s); thus there are $3^{n-1} - b_{n-1}$ sequences ending in 0. So, by the addition principle,

$$b_n = b_{n-1} + b_{n-1} + 3^{n-1} - b_{n-1} \quad \text{i.e. } b_n = b_{n-1} + 3^{n-1}.$$

Again we can find b_n by iteration:

$$b_n = 3^{n-1} + b_{n-1} = 3^{n-1} + (3^{n-2} + b_{n-2}) = \cdots$$

$$= 3^{n-1} + 3^{n-2} + \cdots + 3^1 + b_1.$$

But $b_1 = 1$ (why?), so

$$b_n = 1 + 3 + \cdots + 3^{n-1} = \frac{1}{2}(3^n - 1).$$

Example 2.3 (Paving a garden path)

A path is 2 metres wide and n metres long. It is to be paved using paving stones of size 1m $\times$ 2m. In how many ways can the paving be accomplished?

Solution

Let p_n denote the number of pavings of a $2 \times n$ path. Clearly $p_1 = 1$ since one paving stone fills the path. Also, $p_2 = 2$, the two possibilities being shown in Figure 2.1(a), and $p_3 = 3$ (Figure 2.1(b)).

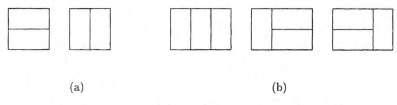

(a)   (b)

Figure 2.1

It might appear that $p_n = n$ for all n, but check now that $p_4 = 5$. What is p_n?
For a $2 \times n$ path, the paving must start with one of the options shown in Figure 2.2.

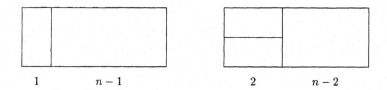

1            $n-1$ 2 $n-2$

Figure 2.2

In the first case it can be completed in p_{n-1} ways; in the second it can be completed in p_{n-2} ways. So, again by the addition principle,

$$p_n = p_{n-1} + p_{n-2} \qquad (n \geq 3).$$

This is a **second order** recurrence relation, since each p_n is given in terms of the previous two. We obtain $p_5 = 5 + 3 = 8$, $p_6 = 8 + 5 = 13, p_7 = 13 + 8 = 21$, etc; the sequence (p_n) thus turns out to be the well-known **Fibonacci sequence** (F_n):

$$1, 2, 3, 5, 8, 13, 21, 34, 55, 89, \ldots.$$

Fibonacci, or Leonardo of Pisa (c. 1200 AD) introduced this sequence when investigating the growth of the rabbit population (see Exercise 2.5); it crops up amazingly frequently in diverse mathematical situations. We shall obtain a formula for F_n in the next section.

Example 2.4 (Flags)

A flag is to consist of n horizontal stripes, where each stripe can be any one of red, white and blue, no two adjacent stripes having the same colour. Under these conditions, the first (top) stripe can be any of three colours, the second has two possibilities, the third has two, and so on (each stripe avoiding the colour of the one above it); so there are $3 \times 2^{n-1}$ possible designs.

Suppose now that, in order to avoid possible confusion of flying the flag upside-down, it is decreed that the top and bottom stripes should be of different colours. Let a_n denote the number of such flags with n stripes. Then $a_1 = 0$ (why?) and $a_2 = 6$. Further, since there is a one-to-one correspondence between flags of n stripes with bottom stripe same as top, and flags of $n-1$ stripes with bottom stripe different from top,

$$
\begin{aligned}
a_n &= 3 \times 2^{n-1} \; - \; \text{(no. of flags with bottom colour same as top colour)} \\
&= 3 \times 2^{n-1} \; - \; \text{(no. of flags of } n-1 \text{ stripes with bottom colour} \\
&\qquad\qquad\qquad \text{different from top).}
\end{aligned}
$$

Thus

$$a_n = 3.2^{n-1} - a_{n-1}. \tag{2.1}$$

We could iterate again (try it!), but here is another method. Since

$$a_n + a_{n-1} = 3.2^{n-1}$$

we also have

$$a_{n-1} + a_{n-2} = 3.2^{n-2}$$

whence

$$2(a_{n-1} + a_{n-2}) = 3.2^{n-1} = a_n + a_{n-1}.$$

Thus

$$a_n = a_{n-1} + 2a_{n-2}. \tag{2.2}$$

This again is a second order recurrence relation; we now show how to solve it.

2.2 The Auxiliary Equation Method

In this section we concentrate on recurrence relations of the form

$$a_n = Aa_{n-1} + Ba_{n-2} \qquad (n \geq 3) \qquad (2.3)$$

where A, B are constants, $B \neq 0$, and where a_1 and a_2 are given. Equation (2.3) is called a second order linear recurrence relation with constant coefficients; it turns out that there is a very neat method of solving such recurrences.

First, we ask: are there any real numbers $\alpha \neq 0$ such that $a_n = \alpha^n$ satisfies (2.3)? Substituting $a_n = \alpha^n$ into (2.3) gives $\alpha^n = A\alpha^{n-1} + B\alpha^{n-2}$, i.e. $\alpha^2 = A\alpha + B$. Thus $a_n = \alpha^n$ is a solution of (2.3) precisely when α is a solution of the **auxiliary equation**

$$x^2 = Ax + B. \qquad (2.4)$$

Thus if α and β are distinct roots of (2.4), $a_n = \alpha^n$ and $a_n = \beta^n$ both satisfy (2.3). If the auxiliary equation has a repeated root α, then

$$x^2 - Ax - B = (x - \alpha)^2 = x^2 - 2\alpha x + \alpha^2$$

so that $A = 2\alpha$ and $B = -\alpha^2$. In this case $a_n = n\alpha^n$ also satisfies (2.3), since

$$Aa_{n-1} + Ba_{n-2} = A(n-1)\alpha^{n-1} + B(n-2)\alpha^{n-2}$$

$$= 2(n-1)\alpha^n - (n-2)\alpha^n = n\alpha^n = a_n.$$

We now prove

Theorem 2.1

Suppose (a_n) satisfies (2.3), and that a_1 and a_2 are given. Let α, β be the roots of the auxiliary equation (2.4). Then

(i) if $\alpha \neq \beta$, there are constants K_1, K_2 such that $a_n = K_1\alpha^n + K_2\beta^n$ for all $n \geq 1$;

(ii) if $\alpha = \beta$, there are constants K_3, K_4 such that $a_n = (K_3 + nK_4)\alpha^n$ for all $n \geq 1$.

Proof

(i) Choose K_1, K_2 so that $a_1 = K_1\alpha + K_2\beta, a_2 = K_1\alpha^2 + K_2\beta^2$, i.e. take

$$K_1 = \frac{a_1\beta - a_2}{\alpha(\beta - \alpha)}, \quad K_2 = \frac{a_1\alpha - a_2}{\beta(\alpha - \beta)}. \qquad (2.5)$$

Then the assertion that $a_n = K_1\alpha^n + K_2\beta^n$ is certainly true for $n = 1, 2$. We now proceed by induction. Assume the assertion is true for all $n \leq k$. Then

$$a_{k+1} = Aa_k + Ba_{k-1} = A(K_1\alpha^k + K_2\beta^k) + B(K_1\alpha^{k-1} + K_2\beta^{k-1})$$

$$= K_1\alpha^{k-1}(A\alpha + B) + K_2\beta^{k-1}(A\beta + B)$$

$$= K_1\alpha^{k+1} + K_2\beta^{k+1},$$

so the result follows.

(ii) Choose K_3, K_4 so that $A_1 = (K_3 + K_4)\alpha$, $a_2 = (K_3 + 2K_4)\alpha^2$, i.e. take

$$K_3 = \frac{2a_1\alpha - a_2}{\alpha^2}, \quad K_4 = \frac{a_2 - a_1\alpha}{\alpha^2}. \tag{2.6}$$

Then the assertion that $a_n = (K_3 + nK_4)\alpha^n$ is certainly true for $n = 1, 2$. Assume it is true for all $n \leq k$. Then

$$a_{k+1} = Aa_k + Ba_{k-1} = A(K_3 + kK_4)\alpha^k + B(K_3 + (k-1)K_4)\alpha^{k-1}$$

$$= K_3\alpha^{k-1}(A\alpha + B) + K_4\alpha^{k-1}(Ak\alpha + B(k-1))$$

$$= K_3\alpha^{k+1} + K_4\alpha^{k-1}(2k - \alpha^2(k-1))$$

$$= K_3\alpha^{k+1} + K_4(k+1)\alpha^{k+1},$$

as required.

Example 2.4 (continued)

In the flag problem we obtained the recurrence relation $a_n = a_{n-1} + 2a_{n-2}$, where $a_1 = 0, a_2 = 6$. The auxiliary equation $x^2 - x - 2 = 0$ has solutions $\alpha = -1, \beta = 2$, so

$$a_n = K_1(-1)^n + K_2 2^n$$

where $0 = -K_1 + 2K_2$ and $6 = K_1 + 4K_2$, i.e. $K_1 = 2, K_2 = 1$. So

$$a_n = 2(-1)^n + 2^n.$$

Example 2.3 (continued)

The **Fibonacci sequence** (F_n) is given by

$$F_1 = 1, \quad F_2 = 2, \quad F_n = F_{n-1} + F_{n-2} \quad (n \geq 3).$$

The auxiliary equation $x^2 - x - 1 = 0$ has solutions $\frac{1}{2}(1 \pm \sqrt{5})$, so

$$F_n = K_1\alpha^n + K_2\beta^n$$

where $\alpha = \frac{1}{2}(1 + \sqrt{5})$, $\beta = \frac{1}{2}(1 - \sqrt{5})$. The initial condition $F_1 = 1, F_2 = 2$, along with (2.5), yield $K_1 = \frac{\alpha}{\sqrt{5}}$, $K_2 = \frac{-\beta}{\sqrt{5}}$, so that

$$F_n = \frac{1}{\sqrt{5}}\alpha^{n+1} - \frac{1}{\sqrt{5}}\beta^{n+1} = \frac{1}{\sqrt{5}}\left(\frac{1+\sqrt{5}}{2}\right)^{n+1} - \frac{1}{\sqrt{5}}\left(\frac{1-\sqrt{5}}{2}\right)^{n+1}. \quad (2.7)$$

This result may seem rather odd since F_n is to be an integer. Check that expansion by the binomial theorem leads to a cancellation of all terms involving $\sqrt{5}$, giving

$$F_n = \frac{1}{2^n}\left\{ \binom{n+1}{1} + 5\binom{n+1}{3} + 5^2\binom{n+1}{5} + \cdots \right\}.$$

This again is a surprise since it is by no means obvious that the sum of binomial coefficients should be divisible by 2^n.

Note that, since $|\beta| < 1$, the second term in (2.7) tends to 0 as $n \to \infty$, giving

$$\frac{F_{n+1}}{F_n} \to \frac{1+\sqrt{5}}{2}, \text{ the golden ratio.}$$

Example 2.5

Solve the recurrence relation $a_n = 4a_{n-1} - 4a_{n-2}$ $(n \geq 3), a_1 = 1, a_2 = 3$.

Solution

The auxiliary equation is $x^2 - 4x + 4 = 0$, i.e. $(x - 2)^2 = 0$, so

$$a_n = (K_1 + nK_2)2^n.$$

The initial conditions give $1 = 2(K_1 + K_2)$, $3 = 4(K_1 + 2K_2)$, whence $K_1 = K_2 = \frac{1}{4}$. Thus

$$a_n = (n + 1)2^{n-2}.$$

The auxiliary equation method extends to higher order recurrences in the obvious way.

Example 2.6

Suppose that $a_1 = 3, a_2 = 6, a_3 = 14$ and, for $n \geq 4$,

$$a_n = 6a_{n-1} - 11a_{n-2} + 6a_{n-3}.$$

Then the auxiliary equation is $x^3 - 6x^2 + 11x - 6 = 0$, i.e. $(x-1)(x-2)(x-3) = 0$, so $a_n = K_1 + K_2 2^n + K_3 3^n$. Using the initial conditions, we get

$$a_n = 1 + 2^{n-1} + 3^{n-1}.$$

Non-homogeneous recurrence relations

The auxiliary equation method has been used for recurrence relations such as $a_n = a_{n-1} + 2a_{n-2}$. These are **homogeneous** linear recurrences with constant coefficients: a_n is a linear combination of some of the previous a_i. We now briefly consider the **non-homogeneous** case, e.g.

$$a_n = Aa_{n-1} + Ba_{n-2} + t_n,$$

where t_n is some function of n. One example of this was (2.1), which we solved by manipulating it into a second order homogeneous recurrence; but now we give an alternative method of solution. For we can obtain a solution by first finding the solution of the recurrence relation obtained by replacing t_n by 0, and then adding to it **any** particular solution of the non-homogeneous recurrence.

Example 2.4 (again)

We solve $a_n = -a_{n-1} + 3.2^{n-1}$, $a_1 = 0$.

Solution

First we solve $a_n = -a_{n-1}$. We could use the auxiliary equation $x = -1$, but it is easy just to spot that $a_n = (-1)^{n-1}a_1$, i.e. $a_n = K(-1)^n$. For a particular solution of $a_n = -a_{n-1} + 3.2^{n-1}$, we try something sensible such as $a_n = A2^n$. Substituting gives $A2^n = -A2^{n-1} + 3.2^{n-1}$, whence $A = 1$. So we have $a_n = K(-1)^n + 2^n$. Since $a_1 = 0$, we need $K = 2$; so we have finally $a_n = 2(-1)^n + 2^n$, as before.

Note that the initial conditions are not applied until the final stage of the procedure.

2.3 Generating Functions

The generating function of a sequence $a_1, a_2, a_3, \ldots$ is defined to be

$$f(x) = \sum_{i=1}^{\infty} a_i x^i.$$

For example, the generating function of the Fibonacci sequence is

$$x + 2x^2 + 3x^3 + 5x^4 + \cdots.$$

If a sequence starts with a_0 we take $f(x) = \sum_{i=0}^{\infty} a_i x^i$; for example, the generating function of the sequence $a_n = 2^n$ ($n \geq 0$) is

$$1 + 2x + 2^2 x^2 + \cdots = \frac{1}{1 - 2x}.$$

Sometimes, given a recurrence relation, it is possible to find the generating function of the sequence and then to find a_n by reading off the coefficient of x^n.

Example 2.4 (yet again!)

Consider the recurrence relation $a_n = 3.2^{n-1} - a_{n-1}$ $(n \geq 2)$, $a_1 = 0$. Let $f(x) = a_1 x + a_2 x^2 + \cdots$. Then

$$f(x) = a_1 x + (3.2 - a_1)x^2 + (3.2^2 - a_2)x^3 + \cdots$$

$$= a_1 x + 3(2x^2 + 2^2 x^3 + \cdots) - (a_1 x^2 + a_2 x^3 + \cdots)$$

$$= 0 + 6x^2(1 + 2x + 2^2 x^2 + \cdots) - x f(x).$$

Thus $(1+x)f(x) = \frac{6x^2}{1-2x}$ so that

$$f(x) = 6x^2 \frac{1}{(1+x)(1-2x)} = 2x^2 \left(\frac{2}{1-2x} + \frac{1}{1+x} \right)$$

on using the method of partial fractions. Thus

$$f(x) = 4x^2(1 + 2x + 2^2 x^2 + \cdots) + 2x^2(1 - x + x^2 - \cdots).$$

Reading off the coefficient of x^n gives

$$a_n = 4.2^{n-2} + 2(-1)^{n-2} = 2^n + 2(-1)^n,$$

as before.

Example 2.5 (again)

$a_n = 4a_{n-1} - 4a_{n-2}$ $(n \geq 3)$, $a_1 = 1$, $a_2 = 3$.

$$f(x) = a_1 x + a_2 x^2 + a_3 x^3 + a_4 x^4 + \cdots$$

$$= x + 3x^2 + (4a_2 - 4a_1)x^3 + (4a_3 - 4a_2)x^4 + \cdots$$

$$= x + 3x^2 + 4(a_2 x^3 + a_3 x^4 + \cdots) - 4(a_1 x^3 + a_2 x^4 + \cdots)$$

$$= x + 3x^2 + 4x(f(x) - a_1 x) - 4x^2 f(x),$$

so that

$$(1 - 4x + 4x^2)f(x) = x + 3x^2 - 4x^2 = x - x^2.$$

Thus

$$f(x) = \frac{x - x^2}{(1 - 2x)^2}.$$

Now, since

$$\frac{1}{1-x} = 1 + x + x^2 + \cdots,$$

differentiating gives

$$\frac{1}{(1-x)^2} = 1 + 2x + 3x^2 + \cdots,$$

so that

$$\frac{1}{(1-2x)^2} = 1 + 2.2x + 3.2^2 x^2 + \cdots.$$

Thus

$$f(x) = (x - x^2)(1 + 2.2x + 3.2^2 x^2 + 4.2^3 x^3 + \cdots)$$

whence

$$a_n = \text{coefficient of } x^{n-1} \text{ in } 1 + 2.2x + \cdots$$

$$- \text{ coefficient of } x^{n-2} \text{ in } 1 + 2.2x + \cdots.$$

$$= n.2^{n-1} - (n-1)2^{n-2} = (n+1)2^{n-2}$$

as before.

2.4 Derangements

Suppose that n people at a party leave their coats in the cloakroom. After the party, they each take a coat at random. How likely is it that no person gets the correct coat?

A **derangement** of $1, \ldots, n$ is a permutation π of $1, \ldots, n$ such that $\pi(i) \neq i$ for each i. For example, there are nine derangements of $1, 2, 3, 4$:

$$
\begin{array}{cccc}
2 & 4 & 1 & 3 \\
2 & 1 & 4 & 3 \\
2 & 3 & 4 & 1 \\
3 & 1 & 4 & 2 \\
3 & 4 & 2 & 1 \\
3 & 4 & 1 & 2 \\
4 & 1 & 2 & 3 \\
4 & 3 & 1 & 2 \\
4 & 3 & 2 & 1 \\
\end{array}
$$

In each of these 1 is not in the first place, 2 is not in the second, and so on. Let d_n denote the number of derangements of $1, \ldots, n$. Then (check!)

$$d_1 = 0, \quad d_2 = 1, \quad d_3 = 2, \quad d_4 = 9.$$

Our aim is to obtain a recurrence relation for the d_i and then use it to obtain a formula for d_n. Before proceeding to the recurrence relation, note that d_n is the number of ways of assigning n objects to n boxes, where, for each object, there is one prohibited box, and where each box is prohibited to just one object. Above, the objects and boxes are both labelled by $1, \ldots, n$, with box (position) i prohibited to object (number) i, but the labelling of the boxes and objects is of course arbitrary and does not affect the problem.

Next note that in three of the nine derangements of $1, 2, 3, 4$ listed above, 4 swaps places with another number: this happens in 2143, 3412 and 4321. In the remaining derangements 4 does **not** swap places with another. With this in mind, we put

$$d_n = e_n + f_n$$

where e_n, f_n denote the numbers of derangements of $1, \ldots, n$ in which n swaps, does not swap, places with another. Now if n swaps places with i (and there are $n - 1$ possible choices for i), the remaining $n - 2$ numbers have to be deranged, and this can be done in d_{n-2} ways; so

$$e_n = (n - 1)d_{n-2}.$$

If n does not swap places with any other, then some r goes to place n (and there are $n - 1$ choices of r), while n does not go to place r. So we have to assign places to $1, \ldots, n$, excluding r, where the places available are $1, \ldots, n - 1$, and where each has precisely one forbidden place (for $i \neq r, n$, place i is forbidden; for $i = n$, place r is forbidden). So there are d_{n-1} possible arrangements, and so

$$f_n = (n - 1)d_{n-1}.$$

Thus by the addition principle, we have

$$\boxed{d_n = (n - 1)(d_{n-1} + d_{n-2}).} \qquad (2.8)$$

Using this recurrence we get

$$d_5 = 4(9 + 2) = 44, \qquad d_6 = 5(44 + 9) = 265,$$

and so on.

The recurrence (2.8) does not permit the use of the auxiliary equation method, since the coefficients of d_{n-1} and d_{n-2} are not constants. However, we can manipulate (2.8) into a more manageable form. Equation (2.8) can be rewritten as

$$d_n - nd_{n-1} = -(d_{n-1} - (n - 1)d_{n-2}),$$

where the expression on the right is the negative of the expression on the left, with n replaced by $n - 1$. So iteration gives

$$d_n - nd_{n-1} = -(d_{n-1} - (n - 1)d_{n-2})$$

$$= (-1)^2(d_{n-2} - (n - 2)d_{n-3})$$

$$\vdots$$

$$= (-1)^{n-2}(d_2 - 2d_1) = (-1)^n(1 - 0) = (-1)^n,$$

i.e.

$$\boxed{d_n - nd_{n-1} = (-1)^n.}$$ (2.9)

Thus

$$\frac{d_n}{n!} - \frac{d_{n-1}}{(n-1)!} = \frac{(-1)^n}{n!}.$$

If we now sum the identities

$$\frac{d_m}{m!} - \frac{d_{m-1}}{(m-1)!} = \frac{(-1)^m}{m!}$$

over $m = 2, 3, \ldots, n$, we get cancellations on the left, giving

$$\frac{d_n}{n!} - \frac{d_1}{1!} = \frac{(-1)^2}{2!} + \frac{(-1)^3}{3!} + \cdots + \frac{(-1)^n}{n!} = \sum_{m=2}^{n} \frac{(-1)^m}{m!} = \sum_{m=0}^{n} \frac{(-1)^m}{m!}.$$

But $d_1 = 0$, so we obtain

$$d_n = n! \sum_{m=0}^{n} \frac{(-1)^m}{m!} = n!\{1 - \frac{1}{1!} + \frac{1}{2!} - \cdots + \frac{(-1)^n}{n!}\}.$$ (2.10)

One interesting consequence of (2.10) is that, as $n \to \infty$,

$$\frac{d_n}{n!} \to \frac{1}{e},$$

so the probability of no one getting their own coat back after the party tends to $\frac{1}{e} = 0.367\,88$ as $n \to \infty$. Indeed, for n as small as 6,

$$\frac{d_6}{6!} = \frac{265}{720} = 0.368\,06,$$

agreeing with $\frac{1}{e}$ to 3 decimal places.

Example 2.7

(a) Find the number of permutations of $1, \ldots, n$ in which exactly k of the numbers are in their correct position, and deduce that

$$n! = \sum_{\ell=0}^{n} \binom{n}{\ell} d_\ell.$$ (2.11)

(b) What is the average number of numbers in their correct position in a random permutation of $1, \cdots, n$?

Solution

(a) There are $\binom{n}{k}$ ways of choosing the k numbers to be fixed. The remaining $n - k$ have to be deranged, and this can be done in d_{n-k} ways. So there are $\binom{n}{k}d_{n-k}$ permutations with exactly k fixed numbers.

But any of the $n!$ permutations fixes k numbers for some k between 0 and n. So

$$n! = \sum_{k=0}^{n} \binom{n}{k} d_{n-k} = \sum_{\ell=0}^{n} \binom{n}{\ell} d_\ell$$

on putting $\ell = n - k$.

(b) The average number of fixed numbers in a permutation of $1, \ldots, n$ is

$$\frac{1}{n!} \sum_{k=0}^{n} k \binom{n}{k} d_{n-k} = \frac{1}{n!} \sum_{k=1}^{n} k \binom{n}{k} d_{n-k}$$

$$= \frac{1}{n!} \sum_{k=1}^{n} n \binom{n-1}{k-1} d_{n-k} \text{ by } (1.2)$$

$$= \frac{1}{(n-1)!} \sum_{k=1}^{n} \binom{n-1}{n-k} d_{n-k}$$

$$= \frac{1}{(n-1)!} \sum_{\ell=0}^{n-1} \binom{n-1}{\ell} d_\ell \text{ (on putting } \ell = n - k)$$

$$= \frac{1}{(n-1)!} (n-1)! \qquad \text{by}(2.11)$$

$$= 1.$$

So the average number of fixed numbers is 1.

Alternative proofs of (2.10)

A proof of (2.10) using the inclusion-exclusion principle will be given in Chapter 6. Here we give yet another proof, a simple application of the inversion principle, as in Corollary 1.15, applied to (2.11).

In (2.11), put a_n for $n!$ and b_n for d_n. Then (2.11) is

$$a_n = \sum_{k=0}^{n} \binom{n}{k} b_k,$$

so that, by Corollary 1.15,

$$d_n = \sum_{k=0}^{n} (-1)^{n+k} \binom{n}{k} k! = \sum_{k=0}^{n} (-1)^{n+k} \frac{n!}{(n-k)!}$$

$$= n! \sum_{\ell=0}^{n} \frac{(-1)^{2n-\ell}}{\ell!} \quad \text{(on putting } \ell = n - k)$$

$$= n! \sum_{\ell=0}^{n} \frac{(-1)^{\ell}}{\ell!}.$$

2.5 Sorting Algorithms

Given a pile of exam scripts, we might want to **sort** them, i.e. put them in increasing or decreasing order of marks. Are there any efficient ways of doing this? We start with a simple but not very efficient procedure.

Bubblesort

Take a list of n numbers, in random order. Compare the first two, swapping them round if they are not in increasing order. Then compare the second and third numbers, again swapping if necessary. In this way proceed up the sequence; the largest number will then be at the end. Next repeat the whole process for the first $n - 1$ numbers: this will take the second largest to the second last position. Repeat for the first $n - 2$, and so on.

The total number of comparisons involved in this procedure is

$$(n - 1) + (n - 2) + \cdots + 2 + 1 = \frac{1}{2}n(n - 1) = \frac{1}{2}n^2 - \frac{1}{2}n,$$

so we say that the bubblesort algorithm has $O(n^2)$ complexity.

Example 2.8

Start with $7, 10, 4, 6, 3$.
   After the first 4 comparisons we have $7, 4, 6, 3, 10$.
After the next 3 comparisons we have $4, 6, 3, 7, 10$.
After the next 2, we have $4, 3, 6, 7, 10$.
After the final comparison we have $3, 4, 6, 7, 10$.

Mergesort

The idea here is to split the given list into two (roughly) equal parts, sort each separately, and then merge (combine) them.

The process of combining two sorted lists of lengths ℓ and m into one list can be accomplished by $\ell + m - 1$ comparisons. For suppose we have two such lists, both in increasing order. Compare the first (smallest) numbers in the lists, and take the smaller as the first member of a new list L, crossing it out of its original position. Repeat the process to find the second member of L, and so on. The number of comparisons is clearly $\ell + m - 1$, since when only one number from the two original lists is left no comparison is necessary.

Before the merging takes place, the two halves of the original list can be sorted by a similar method. Let t_n denote the number of comparisons needed to sort a list of n members by this method. If we split n into $\ell + k$, then $t_n = t_\ell + t_k + \ell + k - 1 = t_\ell + t_k + n - 1$.

Thus, if we consider the particular case where $n = 2^m$, so that the lists can be bisected at each stage, we have

$$t_{2^m} = 2t_{2^{m-1}} + (2^m - 1).$$

Put $a_m = t_{2^m}$; then the recurrence relation becomes

$$a_m = 2a_{m-1} + (2^m - 1). \tag{2.12}$$

Using the method of Section 2.2, first solve the homogeneous recurrence $a_m = 2a_{m-1}$. The solution is clearly $a_n = A2^n$ for some constant A. We then have to find a particular solution of (2.12). Try

$$a_n = Bn2^n + C.$$

(Trying $a_n = B.2^n + C$ would not work, since $a_n = 2^n$ is already a solution of the homogeneous recurrence; so we take the hint given by Theorem 2.1(ii) and insert n.) We then require

$$Bn2^n + C = 2B(n - 1)2^{n-1} + 2C + 2^n - 1$$

i.e.

$$0 = -B.2^n + 2^n - 1 + C.$$

So take $B = C = 1$ to obtain finally $a_n = A.2^n + n2^n + 1$. But $a_1 = 1$, so $A = -1$, giving

$$a_n = 2^n(n - 1) + 1.$$

Thus $t_{2^m} = 1 + 2^m(m - 1)$. On putting $n = 2^m$, we get

$$t_n = 1 + n(\log_2 n - 1),$$

so the mergesort method has complexity $O(n \log n)$, an improvement on the $O(n^2)$ of bubblesort.

2.6 Catalan Numbers

In this section we introduce a well-known sequence of numbers known as the Catalan numbers, which arise as the counting numbers of a remarkable number of different types of structure. They are named after the Belgian mathematician E.C. Catalan (1814–1894) who discussed them in his publications, but they had been studied earlier by several mathematicians, including Euler in his work on triangulating polygons (to be discussed shortly).

We describe fully one of the occurrences of Catalan numbers, and begin with the following easy problem.

Example 2.9

How many "up-right" routes are there from A to B in Figure 2.3?

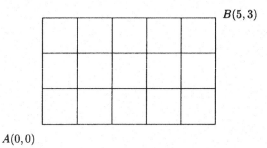

Figure 2.3

Solution

By an "up-right" route we mean a path from A to B following edges of squares, always moving up or to the right. Any path must consist of 8 moves, 5 of which must be to the right, and 3 up. So the total number of possible routes is $\binom{8}{3}$.

More generally, the number of up-right routes from the bottom left vertex to the top right vertex of an $m \times n$ array is $\binom{m+n}{n}$.

Suppose we now have a **square** $n \times n$ array, and ask for the number p_n of up-right paths from bottom left to top right **which never go above the diagonal** AB. In the case $n = 3$, shown in Figure 2.4, there are 5 such routes represented by $RURURU$, $RURRUU$, $RRUURU$, $RRURUU$, $RRRUUU$ where R, U stand respectively for right, up. Thus $p_3 = 5$. What is p_n?

Any qualifying route (let's call it a **good** route) from A to B must "hit" the diagonal at some stage before B, even if it is only at A. So consider any good route from A to B, and suppose that, prior to reaching B, it **last** met the diagonal at the point $C(m,m)$ where $1 \leq m < n$. Then there are p_m possibilities for the part of the route between A and C. The route must then proceed to $D(m+1, m)$, and eventually to $E(n, n-1)$, but it must **never** go

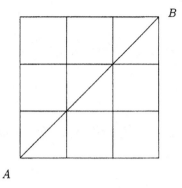

Figure 2.4

above the line DE, since otherwise C would not have been the last hit before B. But D and E are opposite vertices of a square of side length $n - m - 1$, so there are p_{n-m-1} good routes from D to E. See Figure 2.5.

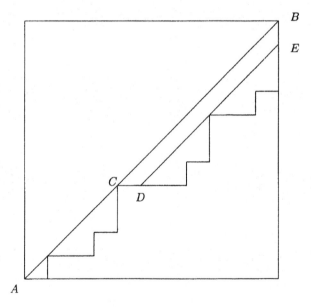

Figure 2.5

By the multiplication principle, the number of good routes from A to B, with (m, m) as the last contact with the diagonal before B, is therefore $p_m p_{n-m-1}$.

Since m can take any value from 0 to $n - 1$, it now follows from the addition principle that, with $p_0 = 1$,

$$p_n = \sum_{m=0}^{n-1} p_m p_{n-m-1}. \tag{2.13}$$

This recurrence relation differs from the ones met so far, but we can use generating functions to solve it. Let $f(x)$ be the generating function:

$$f(x) = p_0 + p_1 x + p_2 x^2 + \cdots .$$

Then

$$
\begin{aligned}
f^2(x) &= (p_0 + p_1 x + p_2 x^2 + \cdots)(p_0 + p_1 x + p_2 x^2 + \cdots) \\
&= \sum_{n=0}^{\infty} x^n (p_0 p_n + p_1 p_{n-1} + \cdots + p_n p_0) \\
&= \sum_{n=0}^{\infty} p_{n+1} x^n \qquad \text{by (2.13)}.
\end{aligned}
$$

Thus $x f^2(x) = \sum_{n=0}^{\infty} p_{n+1} x^{n+1} = f(x) - 1$, whence

$$x f^2(x) - f(x) + 1 = 0.$$

Solving this quadratic, we obtain

$$f(x) = \frac{1 \pm \sqrt{1 - 4x}}{2x} = \frac{1}{2x} \{ 1 - (1 - 4x)^{\frac{1}{2}} \}.$$

We have to take the negative sign to avoid having a term of the form $\dfrac{1}{x}$ in $f(x)$. So

$$
\begin{aligned}
f(x) &= \frac{1}{2x} \{ 1 - (1 - \frac{1}{2} \cdot 4x - \frac{1}{2} \cdot \frac{1}{2} \cdot \frac{4^2 x^2}{2!} - \frac{1}{2} \cdot \frac{1}{2} \cdot \frac{3}{2} \cdot \frac{4^3 x^3}{3!} - \cdots) \} \\
&= \frac{1}{2x} \{ \frac{1}{2} \cdot 4x + \frac{1}{2} \cdot \frac{1}{2} \cdot \frac{4^2 x^2}{2!} + \frac{1}{2} \cdot \frac{1}{2} \cdot \frac{3}{2} \cdot \frac{4^3 x^3}{3!} + \cdots \} \\
&= 1 + \frac{1}{2} \cdot \frac{4x}{2!} + \frac{1}{2} \cdot \frac{3}{2} \cdot \frac{4^2 x^2}{3!} + \frac{1}{2} \cdot \frac{3}{2} \cdot \frac{5}{2} \cdot \frac{4^3 x^3}{4!} + \cdots .
\end{aligned}
$$

Thus, for $n \geq 1$,

$$
\begin{aligned}
p_n &= \frac{1.3.5. \ldots (2n-1)}{2^n (n+1)!} 4^n = \frac{2^n}{(n+1)!} . 1.3.5. \ldots (2n-1) \\
&= \frac{2^n}{(n+1)!} \cdot \frac{(2n)!}{2^n . n!} = \frac{1}{n+1} \binom{2n}{n}.
\end{aligned}
$$

Thus, for example, $p_3 = \frac{1}{4} \binom{6}{3} = 5$ and $p_4 = \frac{1}{5} \binom{8}{4} = 14$. Note also that $p_0 = 1$ fits in with the convention that $\binom{0}{0} = 1$.

The numbers p_n are the **Catalan numbers**, usually denoted by C_n. Thus

$$\boxed{C_n = \frac{1}{n+1}\binom{2n}{n}.}$$ (2.14)

The sequence $(C_n)_{n\geq 0}$ begins

$$1, 1, 2, 5, 14, 42, 139, 429, \ldots .$$

From (2.13) we have

$$C_m = C_0 C_{m-1} + C_1 C_{m-2} + \cdots + C_{m-1} C_0.$$ (2.15)

As remarked earlier, the Catalan numbers appear in many situations. One immediate interpretation, obtained by replacing R and U by 0 and 1 respectively, is:

C_n = number of binary sequences of length $2n$ containing exactly n 0s and n 1s, such that at each stage in the sequence the number of 1s up to that point never exceeds the number of 0s.

Euler's interest was in the following:

C_{n-2} = number of ways of dividing a convex n-gon into triangles by drawing $n - 3$ non-intersecting diagonals. For example, the C_3 ways of triangulating a pentagon are shown in Figure 2.6.

Figure 2.6

See Exercise 2.16 for this problem and Exercise 2.17 for another appearance of C_n.

Another derivation of the formula (2.14)

We close this section by pointing out that there is an alternative ingenious method of counting good up-right routes, due to D. André (1887). It avoids the rather awkward recurrence relation (2.13), instead making use of a clever mirror principle.

The number of good routes from $A(0,0)$ to $B(n,n)$ which do not cross the diagonal AB is the total number $\binom{2n}{n}$ of up-right routes from A to B minus the number of routes which do cross AB. Let's call routes which cross AB **bad** routes.

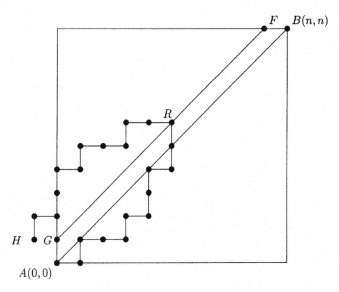

Figure 2.7

Consider any bad route. There will be a **first** point on that route above the diagonal AB; suppose this is the point $R(m, m+1)$. If we replace the part of the route from A to R by its image in the "mirror" GF (see Figure 2.7) then we get an up-right route from $H(-1, 1)$ to $B(n, n)$. Conversely, any up-right route from H to B must cross GF somewhere, and arises from precisely one bad route from A to B. So the number of bad routes is just the number of up-right routes from $(-1, 1)$ to (n, n), which is

$$\binom{n+1+n-1}{n+1} = \binom{2n}{n+1}.$$

So finally the number of good routes from A to B is

$$\binom{2n}{n} - \binom{2n}{n+1} = \binom{2n}{n} - \frac{n}{n+1}\binom{2n}{n} = \frac{1}{n+1}\binom{2n}{n}.$$

Exercises

Exercise 2.1

Solve the recurrence relations
(a) $a_n = \frac{1}{2}a_{n-1} + 1, a_1 = 1$;

(b) $a_n = 5a_{n-1} - 6a_{n-2}, a_1 = -1, a_2 = 1$;

(c) $a_n = 6a_{n-1} - 9a_{n-2}, a_1 = 1, a_2 = 9$;

(d) $a_n = 4a_{n-1} - 3a_{n-2} + 2^n, a_1 = 1, a_2 = 11$.

Exercise 2.2

Let b_n denote the number of n-digit binary sequences containing no two consecutive 0s. Show that $b_n = b_{n-1} + b_{n-2}$ $(n \geq 3)$ and hence find b_n.

Exercise 2.3

Let d_n denote the number of n-digit sequences in which each digit is $0, 1$ or 2, and containing no two consecutive 1s and no two consecutive 2s. Show that $d_n = 2d_{n-1} + d_{n-2}$. Solve this recurrence and deduce that $d_n = 1 + 2\binom{n+1}{2} + 2^2\binom{n+1}{4} + 2^3\binom{n+1}{6} + \cdots$.

Exercise 2.4

Use generating functions to solve Exercise 2.1(a) and 2.1(b).

Exercise 2.5

Fibonacci's rabbits. Start with 1 pair of rabbits, and suppose that each pair produces one new pair in each of the next two generations and then dies. Find f_n, the number of pairs belonging to the nth generation $(f_1 = 1 = f_2)$.

Exercise 2.6

Solve the recurrence relation (2.1) for the flags by iteration.

Exercise 2.7

The Lucas numbers L_n are defined by $L_1 = 1, L_2 = 3, L_n = L_{n-1} + L_{n-2}$ $(n \geq 3)$. Obtain a formula for L_n.

Exercise 2.8

Solve the recurrence (2.12) by using the method given in Example 2.4, first eliminating 1 and then eliminating powers of 2. You should obtain $a_n - 5a_{n-1} + 8a_{n-2} - 4a_{n-3} = 0$.

Exercise 2.9

Verify that if a'_n and a''_n are two solutions of the recurrence $a_n = Aa_{n-1} + Ba_{n-2}$ then $a'_n + a''_n$ is also a solution.

Exercise 2.10

Show that the generating function for the Fibonacci sequence is $\frac{x(1+x)}{1-x-x^2}$. Hence obtain (2.7).

Exercise 2.11

Let $M = \begin{pmatrix} 0 & 1 \\ 1 & 1 \end{pmatrix}$.
(a) Prove that $M^{n+2} = \begin{pmatrix} F_n & F_{n+1} \\ F_{n+1} & F_{n+2} \end{pmatrix}$ where F_n is the nth Fibonacci number.
(b) By taking determinants show that $F_n F_{n+2} - F_{n+1}^2 = (-1)^n$.
(c) By considering the identity $M^{m+n+2} = M^{m+1} M^{n+1}$, prove that $F_{m+n} = F_m F_n + F_{m-1} F_{n-1}$.

Exercise 2.12

Prove that $F_1 + F_2 + \cdots + F_n = F_{n+2} - 2$.

Exercise 2.13

For each of the following, work out the values for the first few values of n and make a guess at the general case. Then prove your guesses by induction.
(a) $F_1 + F_3 + F_5 + \cdots + F_{2n-1}$;
(b) $F_2 + F_4 + F_6 + \cdots + F_{2n}$;
(c) $F_1 - F_2 + F_3 - \cdots + (-1)^{n-1} F_n$.

Exercise 2.14

In bellringing, successive permutations of n bells are played one after the other. Following one permutation π, the next permutation must be obtained from π by moving the position of each bell by at most one place. For example, for $n = 4$, the permutation 1234 could be followed by any one of $2134, 2143, 1324, 1243$. Show that if a_n denotes the number of permutations which could follow $12\ldots n$, then $a_n = a_{n-1} + a_{n-2} + 1$. Hence find a_n.

Exercise 2.15

(a) Let g_n denote the number of subsets of $\{1,\ldots,n\}$ containing no two consecutive integers. Thus, for example, $g_1 = 2$ (include the empty set!) and $g_2 = 3$. Find a recurrence relation for g_n, and deduce that $g_n = F_{n+1}$.

(b) A k-element subset of $\{1,\ldots,n\}$ can be considered as a binary sequence of length n containing k 1s and $n - k$ 0s (see Example 1.12). Use Example 1.17 to show that the number of k-subsets of $\{1,\ldots,n\}$ containing no two consecutive integers is $\binom{n-k+1}{k}$.

(c) Deduce that $F_n = \sum_{k \leq \frac{1}{2}n} \binom{n-k}{k}$. How does this relation show up in Pascal's triangle?

Exercise 2.16

Let t_n denote the number of ways of triangulating a convex $(n + 2)$-gon by drawing $n - 1$ diagonals. Show that $t_n = C_n$ as follows. Label the vertices $1,\ldots,n + 2$, and consider the triangle containing edge 12. If it contains vertex r as its third vertex, in how many ways can the remaining two parts of the interior of the $(n+2)$-gon be triangulated? Deduce that $t_n = \sum t_i t_j$ where summation is over all pairs i, j with $i + j = n - 1$.

Exercise 2.17

Show that if $2n$ points are marked on the circumference of a circle and if a_n is the number of ways of joining them in pairs by n non-intersecting chords, then $a_n = C_n$.

Exercise 2.18

Derive Euler's formula $C_n = \frac{2.4.6\ldots(4n-2)}{(n+1)!}$ for the Catalan numbers, and note that $(n + 1)C_n = (4n - 2)C_{n-1}$.

Exercise 2.19

Prove that $d_n > (n - 1)!$ for all $n \geq 4$.

Exercise 2.20

Insertionsort. Sort a list $x_1,\ldots,x_n$ into increasing order as follows. At stage 1, form list L_1 consisting of just x_1. At stage 2, compare x_1 with x_2 and form list L_2 consisting of x_1 and x_2 in increasing order. At stage

i, when $x_1, \ldots, x_{i-1}$ have been put into list L_{i-1} in increasing order, compare x_i with each x_j in L_{i-1} in turn until its correct position is obtained; this creates list L_i. Repeat until L_n is obtained. Compare the efficiency of this method with that of bubblesort.

Exercise 2.21

In a mathematical model of the population of foxes and rabbits, the populations x_n and y_n of foxes and rabbits at the end of n years are related by $\binom{x_{n+1}}{y_{n+1}} = \begin{pmatrix} 0.6 & 0.5 \\ -0.16 & 1.2 \end{pmatrix} \binom{x_n}{y_n}$.

Show that $5x_{n+2} - 9x_{n+1} + 4x_n = 0$, and hence find x_n in terms of x_0 and y_0.

Deduce that $x_n \to \frac{5}{2}y_0 - x_0$ as $n \to \infty$, provided $x_0 < \frac{5}{2}y_0$. What happens to y_n?

Exercise 2.22

In a football competition, there are n qualifying leagues. At the next stage of the competition, each winner of a league plays a runner up in another league. In how many ways can the winners and the runners up be paired?

3

Introduction to Graphs

We introduce the idea of a graph via some examples, and concentrate on two types of graph, namely trees and planar graphs. Further graph-theoretic topics will be covered in the next chapter.

3.1 The Concept of a Graph

Example 3.1 (The seven bridges of Königsberg)

In the early eighteenth century there were seven bridges over the River Pregel in the Eastern Prussian town of Königsberg (now Kaliningrad). It is said that the residents tried to set out from home, cross every bridge exactly once and return home. They began to believe the task was impossible, so they asked Euler if it were possible. Euler's proof that it was impossible is often taken to be the beginning of the theory of graphs. What Euler essentially did (although his argument was in words rather than pictures) was to reduce the complexity of Figure 3.1(a) to the simple diagram of 3.1(b), where each land mass is represented by a point (vertex) and each bridge by a line (edge). If the desired walk existed, then each time a vertex was visited by using one edge, then another edge would be used up leaving the vertex; so every vertex would have to have an even number of edges incident with it. Since this is not the case, the desired walk is impossible.

The diagram of Figure 3.1(b) is an example of a graph. It has four vertices and seven edges.

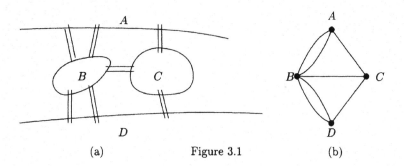

(a)   Figure 3.1   (b)

Example 3.2 (The utilities problem)

An old problem concerns three houses A, B, C which are to be joined to each of the three utilities, gas, water and electricity, without any two connections crossing each other. In other words, can the diagram of Figure 3.2 be redrawn so that no two lines cross? The diagram is another example of a graph.

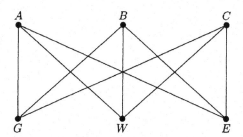

Figure 3.2 The utilities graph

Definition 3.1

A **graph** G consists of a finite set V of **vertices** and a collection E of pairs of vertices called **edges**. The vertices are represented by points, and the edges by lines (not necessarily straight) joining pairs of points. If an edge e joins vertices x and y then x and y are **adjacent** and e is **incident** with both x and y. Any edge joining a vertex x to itself is called a **loop**.

Note that we say E is a collection of pairs, not a set of pairs. This is to allow repeated edges. If two or more edges join the same two vertices, they are called **multiple edges**. For example, the graph of Figure 3.1(b) has two pairs of multiple edges. The graph of the utilities problem is **simple**, i.e. it has no loops or multiple edges.

The number of edges incident with a vertex v in a graph without loops is called the **degree** or **valency** of v and is denoted by $d(v)$. The second name recalls one of the early occurrences of graphs, as drawings of chemical molecules. For example, ethane (C_2H_6) can be represented by the graph of

Figure 3.3, where the two "inside" vertices, of valency 4, represent the two carbon atoms (carbon has valency 4), and the six other vertices, of valency 1, represent hydrogen atoms. Vertices of degree 1 are called **pendant** or **end** vertices.

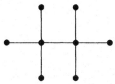

Figure 3.3 Ethane

When a graph contains a loop, the loop is considered to contribute twice to the degree of its incident vertex. This convention enables us to establish the following useful result.

Theorem 3.1

The sum of the degrees of the vertices of a graph is twice the number of edges.

Proof

Each edge contributes twice to the sum of the degrees, once at each end.

This result is sometimes called the **handshaking lemma**: at a party, the total number of hands shaken is twice the number of handshakes. It has an immediate corollary.

Corollary 3.2

In any graph, the sum of the vertex degrees is even.

Example 3.3

The **complete graph** K_n is the simple graph with n vertices, in which each pair of vertices are adjacent. Since each of the n vertices must have degree $n-1$, the number q of edges must satisfy $2q = n(n-1)$, so that $q = \frac{1}{2}n(n-1)$. This of course is as expected, since q is just the number of ways of choosing two of the n vertices, i.e. $q = \binom{n}{2} = \frac{1}{2}n(n-1)$.

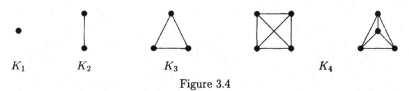

K_1 K_2 K_3 K_4

Figure 3.4

The graphs $K_n, n \leq 4$, are shown in Figure 3.4. The notation K_n is in honour of the Polish mathematician K. Kuratowski (1896–1980) whose important theorem on planarity will be mentioned in Section 3.6. Note that K_4 contains K_3 within it; this idea of one graph being contained in another is formalised in the next definition.

Definition 3.2

A graph H is said to be a **subgraph** of a graph G if the vertex set of H is a subset of the vertex set of G, and the edge set of H is a subset of the edge set of G.

Thus, for example, K_m is a subgraph of K_n wherever $m < n$; simply restrict K_n to m of its vertices.

Finally in this section, we establish some standard notation. From now on, we shall use p and q to denote the numbers of vertices and edges respectively, and by a (p, q)-graph we shall mean a graph with p vertices and q edges. Thus, for example, K_4 is a $(4, 6)$-graph.

3.2 Paths in Graphs

Many important applications of graph theory involve travelling round the graph, in the sense of moving from vertex to vertex along incident edges. We make some definitions related to this idea.

Definition 3.3

A **walk** in a graph G is a sequence of edges of the form

$$v_0 v_1, \ v_1 v_2, \ v_2 v_3, \ldots, v_{n-1} v_n.$$

This walk is sometimes, in a simple graph, represented more compactly by $v_0 \to v_1 \to v_2 \to \cdots \to v_n$. Note that there is an implied direction to the walk. v_0 is called the **initial** vertex and v_n the **final** vertex of the walk; the number (n) of edges is called the **length** of the walk.

A walk in which all the edges are distinct is called a **trail**. A trail in which all vertices $v_0, \ldots, v_n$ are distinct (except possibly $v_n = v_0$) is called a **path**; a path $v_0 \to \ldots \to v_n$ with $v_n = v_0$ is called a **cycle**.

Example 3.4

In the graph of Figure 3.5,

$z \to u \to y \to v \to u$	is a trail but not a path;
$u \to y \to w \to v$	is a path of length 3;
$u \to y \to w \to v \to u$	is a cycle of length 4.

It seems natural to consider the cycles $u \to y \to v \to u$ and $y \to v \to u \to y$

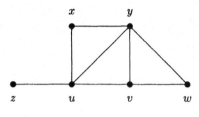

Figure 3.5

to be the same; so often we identify a cycle with the set of its edges. We use the notation (for $n > 1$)

$$C_n = \text{cycle of length } n \text{ (i.e. with } n \text{ edges and vertices)};$$

$$P_n = \text{path of length } n - 1 \text{ (i.e. with } n \text{ vertices)}.$$

Thus, for example, $P_2 = K_2$ and $C_3 = K_3$.

Definition 3.4

A graph is **connected** if, for each pair x, y of vertices, there is a path from x to y. A graph which is not connected is made up of a number of connected pieces, called **components**.

3.3 Trees

Definition 3.5

A **tree** is a connected simple graph with no cycles.

For example, the ethane graph in Figure 3.3 is a tree, as is each P_n. Note that the ethane graph has $p = 8$ and $q = 7$, while P_n has $p = n$ and $q = n - 1$; in each case, $p - q = 1$. This property in fact characterises those connected graphs which are trees. Our proof of this depends upon the following useful result.

Theorem 3.3

If T is a tree with $p \geq 2$ vertices then T contains at least two pendant vertices.

Proof

Since T has p vertices, all paths in T must have length less than p. So there must be a longest path in T, say $v_1 \to v_2 \to \ldots \to v_r$. We claim that v_1 and

v_r both have degree 1. Suppose v_1 has degree > 1; then there is another edge from v_1, say $v_1 v_0$, where v_0 is none of $v_2, \ldots, v_r$ (otherwise there would be a cycle), so $v_0 \to v_1 \to \ldots \to v_r$ would be a longer path. So v_1 has degree 1, and a similar argument holds for v_r.

Theorem 3.4

Let T be a simple graph with p vertices. Then the following statements are equivalent:

(i) T is a tree;

(ii) T has $p - 1$ edges and no cycles;

(iii) T has $p - 1$ edges and is connected.

Proof

(i) $\Rightarrow$ (ii) We have to show that all trees with p vertices have $p - 1$ edges. This is certainly true when $p = 1$. Suppose it is true for all trees with $k \geq 1$ vertices, and let T be a tree with $k + 1$ vertices. Then, by Theorem 3.3, T has an end vertex w. Remove w and its incident edge from T to obtain a tree T' with k vertices. By the induction hypothesis, T' has $k - 1$ edges; so T has $(k - 1) + 1 = k$ edges as required.

(ii) $\Rightarrow$ (iii) Suppose T has $p - 1$ edges and no cycles, and suppose it consists of $t \geq 1$ components, $T_1, \ldots, T_t$, each of which has no cycles and hence must be a tree. Let p_i denote the number of vertices in T_i. Then $\sum_i p_i = p$, and the number of edges in T is $\sum_i (p_i - 1) = p - t$. So $p - t = p - 1$, i.e. $t = 1$, so that T is connected.

(iii) $\Rightarrow$ (i) Suppose T is connected with $p - 1$ edges, but is not a tree. Then T must have a cycle. Removing an edge from a cycle does not destroy connectedness, so we can remove edges from cycles until no cycles are left, preserving connectedness. The resulting graph must be a tree, with p vertices and $q < p-1$ edges, contradicting (ii).

This theorem can be used to establish the tree-like nature of certain chemical molecules.

Example 3.5

Show that the alkanes (paraffins) $C_n H_{2n+2}$ have tree-like molecules.

Solution

Each molecule is represented by a graph with $n + (2n + 2) = 3n + 2$ vertices. Of these, n have degree 4 and $2n + 2$ have degree 1, so, by Theorem 3.1,

$$2q = 4n + 2n + 2 = 6n + 2$$

whence $q = 3n + 1 = p - 1$. Since molecules are connected, the graphs must be trees, by Theorem 3.4.

The first few alkanes are shown in Figure 3.6.

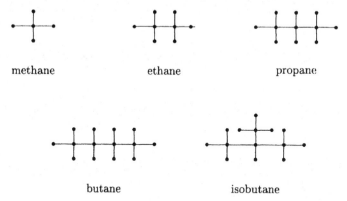

Figure 3.6 Alkanes

Note that there are two "different" trees corresponding to C_4H_{10}.

Definition 3.6

Two graphs G_1, G_2 are **isomorphic** if it is possible to label the vertices of both graphs by the same labels, so that, for each pair u, v of labels, the number of edges joining vertices u and v in G_1 is equal to the number of edges joining u and v in G_2.

Example 3.6

(i) The graphs portrayed by the last two diagrams in Figure 3.4 are isomorphic. (ii) The butane and isobutane graphs (Figure 3.6) are not isomorphic. The second graph has one vertex of degree 4 joined to all the other vertices of degree 4, but this does not happen in the first graph.

Tree diagrams such as those in Figure 3.6 were introduced in 1864 by the chemist A. Crum Brown in his study of isomerism, the occurrence of molecules with the same chemical formula but different chemical properties. The problem of enumerating the non-isomorphic molecules C_nH_{2n+2} was eventually solved by Cayley in 1875, but his solution is beyond the scope of this book.

A related problem was: find $T(n)$, the number of non-isomorphic trees with n vertices. We have $T(1) = T(2) = T(3) = 1$, and, as the reader should check, $T(4) = 2, T(5) = 3, T(6) = 6$. No simple formula for $T(n)$ exists, although $T(n)$ is the coefficient of x^n in a known but very complicated series. However, there does exist a very nice formula for the number of trees on n given labelled vertices. For example, although $T(3) = 1$, there are three **labelled** trees with

vertices labelled $1, 2, 3$ as shown in Figure 3.7. It was established by Cayley in
1889 that the number of labelled trees on n vertices is n^{n-2}. A proof of this
will be given in Chapter 6.

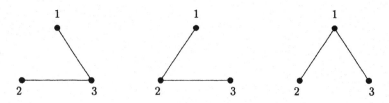

Figure 3.7 Labelled trees.

3.4 Spanning Trees

Suppose that a connected graph represents a railway system, the vertices representing the towns and the edges the railtracks. Suppose also that the gov-
ernment wishes to get rid of as much track as possible, nevertheless retaining
a rail system which connects all the towns. What is required is a tree which is
a subgraph of the given graph, containing all the vertices.

Definition 3.7

A **spanning tree** of a connected graph G is a tree which contains all the
vertices of G and which is a subgraph of G.

Example 3.7

(i) K_3 has three spanning trees, as shown in Figure 3.7.

(ii) K_4 has $16 = 4^2$ spanning trees. Draw them. Do you see how this relates to
     Cayley's 1889 result?

(iii) In the graph of Figure 3.5, the edges zu, xu, uy, yv, yw form a spanning
      tree.

In the case of a **weighted** graph G, i.e. when each edge e of G has a weight $w(e)$
assigned to it, where $w(e)$ is a positive number such as the length of e, then it
may be desired to find a spanning tree of smallest possible total weight. There
are several different algorithms which find such a minimum weight spanning
tree of G.

The greedy algorithm

This is often known as Kruskal's algorithm.

Procedure

(i) Choose an edge of smallest weight.

(ii) At each stage, choose from the edges not yet chosen the edge of smallest weight whose inclusion will not create a cycle.

(iii) Continue until a spanning tree is obtained.

(If the given graph has p vertices, the algorithm will terminate after $p-1$ edges have been chosen.)

Example 3.8

Apply the greedy algorithm to the graph of Figure 3.8.

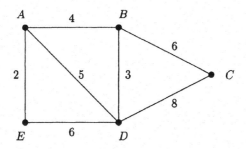

Figure 3.8

Solution

First choose AE (weight 2). Then choose $BD(3)$, then $AB(4)$. We cannot now choose $AD(5)$ since its inclusion would create a cycle $ABDA$. Similarly we cannot choose DE. So choose $BC(6)$. The edges AE, AB, BD, BC then form a minimum weight spanning tree of weight $2 + 3 + 4 + 6 = 15$.

Justification of the greedy algorithm

Suppose that the greedy algorithm produces a tree T, but that there is another spanning tree U which has smaller weight than T. Since $T \neq U$, and both have the same number of edges, there must be an edge in T not in U: let e be such an edge of minimum weight. The addition of e to U must create a cycle C, and this cycle must contain an edge e' which is not in T. Now $w(e') \geq w(e)$, since if $w(e') < w(e)$ then e' would have been chosen by the greedy algorithm rather than e. So if we remove e' from C we obtain a spanning tree V such that $w(V) \leq w(U)$, and V has one more edge in common with T than U had. By repeating this process we eventually change U into T, one edge at a time, and conclude that $w(T) \leq w(U) < w(T)$, a contradiction. So no such U can exist.

The greedy algorithm is so called because it greedily minimises the weight at each step, ignoring possible future complications; fortunately it gets away with this strategy. The disadvantage of the algorithm, however, lies in the difficulty of determining at each stage whether or not a cycle would be created by the inclusion of the smallest weighted edge available (this is particularly true when the graph is large). This problem can be overcome by using a slightly different algorithm, due to Prim (1957). In Prim's algorithm, the graph constructed is connected (and hence a tree) at each stage of the construction (unlike the greedy algorithm, which chose BD immediately after AE in the above example), and at each stage the smallest weight edge is sought which joins the existing tree to a vertex not in the tree. Clearly the inclusion of this edge cannot create a cycle.

Prim's algorithm

(i) Select any vertex, and choose the edge of smallest weight from it.

(ii) At each stage, choose the edge of smallest weight joining a vertex already included to a vertex not yet included.

(iii) Continue until all vertices are included.

Example 3.8 (revisited)

Use Prim's algorithm starting at B. Choose $BD(3)$, then $BA(4)$, then $AE(2)$, then $BC(6)$ to obtain the same spanning tree as before.

A third algorithm operates by **removing** edges from the given graph, destroying cycles, until a spanning tree is left. At each stage remove the largest-weighted edge whose removal does not disconnect the graph. In Example 3.8, we could remove DC, then DE, then AD. Clearly this approach would be quicker than the others if the graph has "few" edges.

3.5 Bipartite Graphs

Definition 3.8

A graph is **bipartite** if its vertex set V can be partitioned into two sets B, W in such a way that every edge of the graph joins a vertex in B to a vertex in W. The partition $V = B \cup W$ is called a **bipartition** of the vertex set.

Example 3.9

Labellings show that the graphs in Figure 3.9 are bipartite. In both graphs, each edge joins a B to a W.

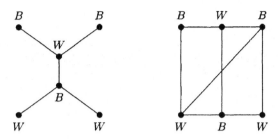

Figure 3.9 Bipartite graphs

If we interpret B and W as black and white, we see that a graph is bipartite precisely when the vertices can be coloured using two colours so that no edge joins two vertices of the same colour. For this reason, bipartite graphs are sometimes called **bichromatic**.

Example 3.10

The cycle C_n is bipartite if and only if n is even.

Theorem 3.5

A connected graph is bipartite if and only if it contains no cycle of odd length.

Proof

If a graph G contains an odd cycle (i.e. a cycle of odd length) then it cannot possibly be bipartite. So suppose now that G contains no odd cycle; we shall show how to colour its vertices B and W.

Choose any vertex v of G, and partition V as $B \cup W$ where

$$B = \{u \in V : \text{ shortest path from } v \text{ to } u \text{ has even length }\},$$

$$W = \{u \in V : \text{ shortest path from } v \text{ to } u \text{ has odd length }\}.$$

We have $u \in B$ since 0 is even; we have to check that no edge of G has both ends in B or both ends in W.

Suppose there is an edge xy with $x \in B$ and $y \in B$. Then, denoting the length of the shortest path from vertex v_1 to vertex v_2 by $d(v_1, v_2)$, we have $d(v, x) = 2m$ and $d(v, y) = 2n$ for some integers m, n. But there is a walk from v to y via x of length $2m + 1$, so $2n \le 2m + 1$. Similarly $2m \le 2n + 1$, so $m = n$.

Denote the shortest paths from v to x and y by $P(x)$ and $P(y)$ respectively. Then, since $m = n$, both $P(x)$ and $P(y)$ have equal lengths. Let w be the **last** vertex on $P(x)$ which is also on $P(y)$ (possibly $w = v$). Then the part of $P(x)$

from w to x and the part of $P(y)$ from w to y must be of equal length, and, since they have only w in common, they must, with edge xy, form an odd cycle. But G has no odd cycles, so the assumption of the existence of the edge xy must be false. So there is no edge with both edges in B; similarly there is no edge with both edges in W.

Corollary 3.6

All trees are bipartite.

Definition 3.9 (Complete bipartite graphs)

A simple bipartite graph with vertex set $V = B \cup W$ is **complete** if every vertex in B is joined to every vertex in W. If $|B| = m$ and $|W| = n$, the graph is denoted by $K_{m,n}$ or by $K_{n,m}$. For example, the utilities graph of Figure 3.2 is $K_{3,3}$, and the methane graph of Figure 3.6 is $K_{1,4}$.

Clearly, $K_{m,n}$ has $m + n$ vertices and mn edges; m of the vertices have degree n, and n of the vertices have degree m.

The complete graphs K_n and the complete bipartite graphs $K_{m,n}$ play important roles in graph theory, particularly in the study of planarity to which we now turn.

3.6 Planarity

A graph is **planar** if it can be drawn in the plane with no edges crossing. The concept of planarity has already appeared in the utilities problem, which can be restated as: is $K_{3,3}$ planar? If a graph is planar, then any drawing of it with no edges crossing is called a **plane** graph. For example, K_4 is planar, as was shown in Figure 3.4; the second drawing of K_4 there was a plane graph, establishing its planarity.

Planar graphs occur naturally in the **four-colour problem**. In colouring a map, it is standard procedure to give adjacent countries different colours. It became apparent that four colour always seemed to be sufficient to colour any map, and a general proof of this statement was attempted by A.B. Kempe in 1879. Ten years later, Heawood discovered that Kempe's "proof" was flawed, and instead of the four-colour theorem we had the four-colour conjecture. Eventually, in 1976, the truth of the conjecture was established by two mathematicians, K. Appel and W. Haken; as the postmark of the University of Illinois asserted, "four colours suffice".

The problem of colouring a map can be transformed into one of colouring the vertices of a planar graph. Given a map, we can represent each region by a vertex, and join two vertices by an edge precisely when the corresponding regions share a common boundary. For example, Figure 3.10 shows a map and a planar graph representing it. So the problem reduces to that of colouring the vertices

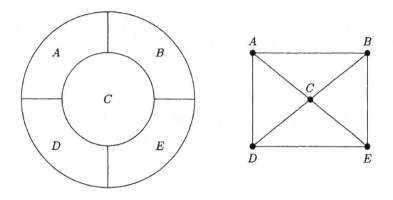

Figure 3.10

of a planar graph with four colours, so that no two adjacent vertices receive the same colour. Colourings of graphs will be discussed further in Chapter 5.

Any plane graph clearly divides the plane into disjoint regions, one of which is infinite. The basic result about plane graphs is known as Euler's formula; Euler initially studied it in the context of polyhedra, and we shall look at this in the next section.

Theorem 3.7 (Euler's formula)

Any connected plane (p, q)-graph divides the plane into r regions, where

$$p - q + r = 2.$$

Proof

If there is a cycle, remove one edge from it. The effect is to reduce q and r by 1 (since two regions are amalgamated into one), and to leave p unchanged. So the resulting graph has $p' = p, q' = q - 1, r' = r - 1$, where $p' - q' + r' = p - q + r$. Repeat this process until no cycles remain. The final graph must be a tree, with $p'' - q'' + r'' = p - (p - 1) + 1 = 2$.

Example 3.11

The plane graph in Figure 3.10 has

$$p - q + r = 5 - 8 + 5 = 2.$$

There are four finite regions and one infinite region.

We now define the **degree** of a region of a plane graph to be the number of encounters with edges in a walk round the boundary of the region.

Example 3.12

In Figure 3.11 regions 3 and 4 have degree 3, the infinite region 1 has degree

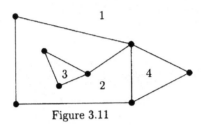

Figure 3.11

5, and region 2 has degree 9 (note that one edge is encountered twice, once on each side).

Parallel to the handshaking lemma we have:

Theorem 3.8

In a connected plane graph, $2q$ = sum of degrees of the regions.

Theorem 3.9

K_n is planar only if $n \leq 4$.

Proof

It is enough to show that K_5 is non-planar. (Why?) Now K_5 has $p = 5, q = 10$, so if a plane drawing of K_5 exists it must have $r = 2 - 5 + 10 = 7$ regions. Each of the seven regions must have degree ≥ 3, so, by Theorem 3.8, $20 = 2q \geq 7 \times 3 = 21$, a contradiction.

Theorem 3.10

$K_{3,3}$ is not planar.

Proof

$K_{3,3}$ has $p = 6$ and $q = 9$, so if a plane drawing exists it must have $r = 2 - 6 + 9 = 5$ regions. Since $K_{3,3}$ is bipartite, with no odd cycles, each region must have degree ≥ 4, so we must have $18 = 2q \geq 4 \times 5 = 20$, a contradiction.

Corollary 3.11

$K_{m,n}$ is planar $\Leftrightarrow$ $\min(m,n) \leq 2$.

The technique of counting the sum of the degrees of the regions is a useful one. We can apply it to the famous Petersen graph, shown in Figure 3.12. (See Section 4.1 and Exercise 5.17 for more about this graph.)

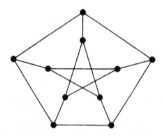

Figure 3.12   The Petersen graph

Example 3.13

The Petersen graph is not planar.

Solution

Suppose a plane drawing exists. Since $p = 10$ and $q = 15$, we would have $r = 2 - 10 + 15 = 7$. Now the shortest cycle in the graph clearly has length 5, so every region must have degree ≥ 5. So we would have a contradiction, $30 = 2q \geq 7 \times 5 = 35$.

Kuratowski's theorem

What makes a graph non-planar? Clearly, if it contains K_5 or $K_{3,3}$ as a subgraph, then it cannot possibly be planar. It was proved in 1930 by the Polish mathematician Kuratowski that, essentially, it is only the presence of a K_5 or a $K_{3,3}$ within a graph that can stop it being planar.

To clarify this statement, we first make the following observation. Since K_5 is not planar, the graph shown in Figure 3.13 cannot be planar either. For if it were, we could make a plane drawing of it, erase b from the edge ac, and obtain a plane drawing of K_5. Inserting a new vertex into an existing edge of a graph is called **subdividing** the edge, and one or more subdivision of edges creates a **subdivision** of the original graph.

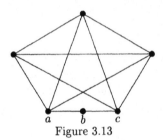

Figure 3.13

Theorem 3.12 (Kuratowski's theorem)

A graph is planar if and only if it does not contain a subdivision of K_5 or $K_{3,3}$ as a subgraph.

The proof of this deep topological result is beyond the scope of this book. But we exhibit the result's usefulness by using it to prove that the Petersen graph is non-planar.

Example 3.13 (again)

In Figure 3.14, Petersen's graph is on the left. On the right is the same graph with two edges removed. This subgraph is a subdivision of $K_{3,3}$ as shown by the labelling of the vertices.

Another test for planarity will be given in Section 4.2.

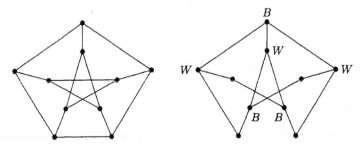

Figure 3.14

Chords of a circle

We close this section on planarity with an application of Euler's formula to a well-known problem concerning chords of a circle.

Suppose we have n points spaced round a circle, and we join each pair of points by a chord, taking care to ensure that no three chords intersect at the same point. Into how many regions is the interior of the circle divided? The cases $n = 3, 4, 5$ are shown in Figure 3.15. It would appear that $n = 6$ should

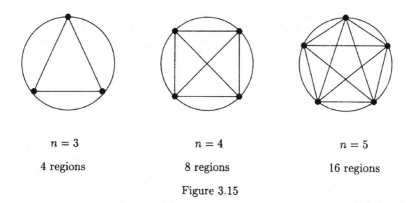

$n = 3$ $n = 4$ $n = 5$

4 regions   8 regions   16 regions

Figure 3.15

give 32 regions. But it does not! (Check!)

Suppose we have n points and have drawn the $\binom{n}{2}$ chords. There will be n regions with a circular arc as a boundary - let's lay them aside and concentrate on the remaining regions. Turn the geometrical picture into a graph by putting a vertex at each of the n given points, and at each crossing point of chords. How many crossing points are there? There is one for each pair of chords which cross. But any pair of crossing chords is obtained by choosing 4 of the given n points and drawing the "cross" chords between them; so there must be $\binom{n}{4}$ crossing points. So the resulting graph has $p = n + \binom{n}{4}$ vertices. Each of the n original vertices has degree $n - 1$, and each of the new $\binom{n}{4}$ vertices has degree 4. So by the handshaking lemma

$$2q = n(n-1) + 4\binom{n}{4}, \qquad \text{i.e. } q = \binom{n}{2} + 2\binom{n}{4}.$$

Thus

$$\begin{aligned} r &= 2 - p + q \\ &= 2 - n - \binom{n}{4} + \binom{n}{2} + 2\binom{n}{4} \\ &= 2 - n + \binom{n}{2} + \binom{n}{4}. \end{aligned}$$

Here r includes a count of 1 for an infinite region, so there are $1 - n + \binom{n}{2} + \binom{n}{4}$ finite regions. We have to add the n boundary regions which we put aside earlier, so finally the number of regions is

$$1 + \binom{n}{2} + \binom{n}{4}.$$

Check that this gives $4, 8, 16$ for $n = 3, 4, 5$, and 31 for $n = 6$.

3.7 Polyhedra

A polyhedron is a solid bounded by a finite number of faces, each of which is polygonal. For example, the pyramid in Figure 3.16(a) is a polyhedron with five vertices, five faces (four triangular, and one square base), and eight edges.

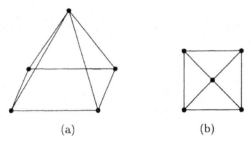

(a)   (b)

Figure 3.16 A pyramid and its plane graph

As was mentioned earlier, Euler's formula arose first in the study of polyhedra, relating the numbers of vertices, faces and edges in a convex polyhedron. (A polyhedron is convex if the straight line segment joining any two of its vertices lies entirely within it.) Such a polyhedron can be represented by a plane graph, obtained by projecting the polyhedron into a plane. The graph in Figure 3.16(b) represents the pyramid; think of the internal vertex as the top of the pyramid, and think of the base of the pyramid as being represented by the infinite region (of degree 4).

The cube is an example of a **regular** polyhedron. A polyhedron is regular if there exist integers $m \geq 3, n \geq 3$ such that each vertex has m faces (or m edges) meeting at it, and each face has n edges on its boundary. For a cube, $m = 3$ and $n = 4$. Convex regular polyhedra are known as **Platonic solids**; they were discussed at great length by the ancient Greeks who knew that there were only 5 such solids. In the next theorem we use the terminology of graphs, moving from a polyhedron to its corresponding graph.

Theorem 3.13

Suppose that a regular polyhedron has each vertex of degree m and each face of degree n. Then (m, n) is one of $(3,3), (3,4), (4,3), (3,5), (5,3)$.

Further, there exist Platonic solids corresponding to each of these pairs.

Proof

We have $p - q + r = 2$, where

$$2q = \text{ sum of vertex degrees } = mp$$

$$\text{and } 2q = \text{ sum of face degrees } = nr.$$

So $(\dfrac{2}{m} - 1 + \dfrac{2}{n})q = 2$, whence

$$(2m + 2n - mn)q = 2mn. \tag{3.1}$$

Thus, trivially, $2m+2n-mn > 0$, i.e. $(m-2)(n-2) < 4$. So $(m-2)(n-2) = 1, 2$ or 3, and the five possibilities arise.

For each possible pair (m, n), we can find q from (3.1) and then deduce the values of p and r. We tabulate these values in Table 3.1, and give the name of the corresponding Platonic solid.

Table 3.1

m	n	q	p	r	Name
3	3	6	4	4	tetrahedron
3	4	12	8	6	cube
4	3	12	6	8	octahedron
3	5	30	20	12	dodecahedron
5	3	30	12	20	icosahedron

Note that the names reflect the number r of faces. The five solids, and their plane graphs, are shown in Figure 3.17.

As well as the five regular polyhedra just discussed, there exist the **semiregular** polyhedra known as the **Archimedean solids**. Although they may well have been known to the Greeks, the first known listing of them is due to Kepler in 1619. These solids have more than one type of face, but they have the property that each vertex has the same pattern of faces around it. For example, the truncated cube, obtained by slicing off each of the eight vertices, has eight triangular faces and six octagonal faces, and, at each vertex, two octagons and one triangle meet.

Example 3.14

A polyhedron is made up of pentagons and hexagons, with three faces meeting at each vertex. Show that there must be exactly 12 pentagonal faces.

Solution

We have $p - q + r = 2$ and $2q = $ sum of vertex degrees $= 3p$. Thus $2q = 6r - 12$. Now suppose there are x pentagonal and y hexagonal faces. Then $r = x + y$ and $2q = $ sum of degrees of faces $= 5x + 6y$. Substituting into $2q = 6r - 12$ gives $5x + 6y = 6x + 6y - 12$, whence $x = 12$.

The case $x = 12, y = 0$ corresponds of course to a dodecahedron. The case $x = 12, y = 20$ corresponds to the pattern often seen on a soccer ball. The corresponding Archimedean solid is a **truncated icosahedron**; the reader

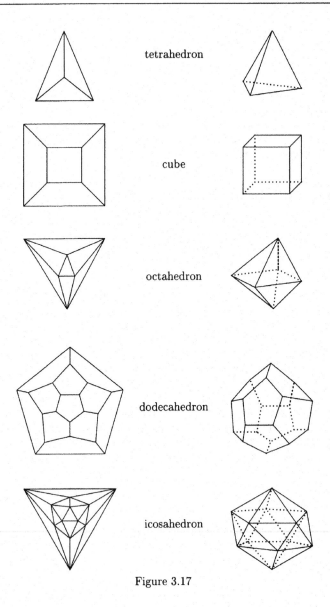

tetrahedron

cube

octahedron

dodecahedron

icosahedron

Figure 3.17

should be able to see how to obtain one by slicing vertices off an icosahedron. This solid aroused great interest in the 1990s when it was discovered that a third form of carbon existed (as well as diamond and graphite).

This form is denoted by C_{60}; the molecular structure is that of 60 carbon atoms situated at the vertices of a truncated icosahedron. The discoverers of this molecule called it **Buckminsterfullerine** (it is commonly known as a Buckyball) since they considered it as similar to a geodesic dome created by the architect R. Buckminster Fuller. But, as we have pointed out, it has been known to mathematicians for a long time.

The graphite form of carbon has the carbon atoms arranged in a flat honeycomb pattern of hexagons. Hexagons tile the plane, so need the addition of n-gons with $n < 6$ to enable a 3-dimensional form to take place. It turns out that 12 pentagons are just right to enable a complete closing up to take place. See Exercise 3.14 for the corresponding problem when pentagons are replaced by squares.

There are other fullerine molecules, such as C_{70} which has 12 pentagons and 25 hexagons; its shape is more like a rugby ball.

Exercises

Exercise 3.1

Prove that the number of vertices of odd degree in a given graph is even.

Exercise 3.2

Show that all alcohols $C_nH_{2n+1}OH$ have tree-like molecules. (The valencies of C, O, H are $4, 2, 1$ respectively.)

Exercise 3.3

Show that if G is a simple graph with p vertices, where each vertex has degree $\geq \frac{1}{2}(p-1)$, then G must be connected. (Hint: how many vertices must each component have?)

Exercise 3.4

How many spanning trees do the graphs in Figure 3.18 have?

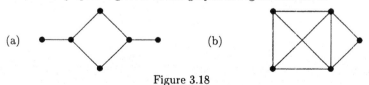

(a)   (b)

Figure 3.18

Exercise 3.5

How many edges must be removed from a connected (p, q)-graph to obtain a spanning tree?

Exercise 3.6

Let $K_{2,3}$ have bipartition $B \cup W$ where $B = \{a, b\}, W = \{x_1, x_2, x_3\}$.
(a) Explain why, in a spanning tree of $K_{2,3}$, there must be precisely one of the vertices x_i joined to both a and b.
(b) How many spanning trees does $K_{2,3}$ have?
(c) How many spanning trees does $K_{2,100}$ have?

Exercise 3.7

Use (a) the greedy algorithm, (b) Prim's algorithm to find a minimum weight spanning tree in the graph shown in Figure 3.19.

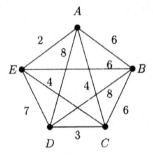

Figure 3.19

Exercise 3.8

The distances between 5 Lanarkshire towns are given in Table 3.2. Find the shortest length of a connecting road network.

Table 3.2

	G	H	A	M	EK
Glasgow	0	10	11	13	9
Hamilton	10	0	8	3	6
Airdrie	11	8	0	8	13
Motherwell	13	3	8	0	8
East Kilbride	9	6	13	8	0

Exercise 3.9

Pan Caledonian Airways (PCA) operates between 12 towns whose coordinates referred to a certain grid are $(0,2), (0,5), (1,0), (1,4), (2,3), (2,4)$, $(3,2), (3,5), (4,4), (4,5), (5,3), (6,1)$. What is the minimum number of flights necessary so that travel by PCA is possible between any two of the towns? Find the minimum total length of such a network of flights.

Exercise 3.10

Determine which of the graphs in Figure 3.20 are planar.

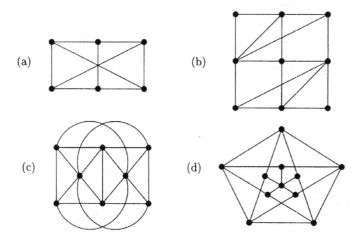

Figure 3.20

Exercise 3.11

A **complete matching** of a graph with $2n$ vertices is a subgraph consisting of n disjoint edges. How many different complete matchings are there in the graph of Figure 3.20(a)?

Exercise 3.12

The graph $G_n (n \geq 1)$ is shown in Figure 3.21.

(a) Is G_n (i) bipartite? (ii) planar?

(b) Let a_n denote the number of complete matchings of G_n. Show that

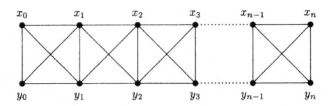

Figure 3.21

$a_1 = 3$ and $a_2 = 5$. Show that $a_n = a_{n-1} + 2a_{n-2}$ $(n \geq 3)$ and hence obtain a formula for a_n.

Exercise 3.13

(a) Show that if G is a simple planar (p, q) graph, $p \geq 3$, then $q \leq 3p - 6$. Deduce that K_5 is not planar.

(b) Show that if G is a simple planar (p, q) graph, $p \geq g$, where g is the **girth** of G, i.e. the length of the shortest cycle in G, then $q \leq \frac{g}{g-2}(p - 2)$.

(c) Deduce from (b) that $K_{3,3}$ and the Petersen graph are both nonplanar.

Exercise 3.14

A convex polyhedron has only square and hexagonal faces. Three faces meet at each vertex. Use Euler's formula to show that there must be exactly six square faces. The cube has no hexagonal faces: give an example with six square faces and at least one hexagonal face. (Try truncating an octahedron.)

Exercise 3.15

Suppose n cuts are made across a pizza. Let p_n denote the maximum number of pieces which can result (this happens when no two cuts are parallel or meet outside the pizza, and no three are concurrent).

Prove that $p_n = \binom{n}{0} + \binom{n}{1} + \binom{n}{2}$.

Exercise 3.16

Let h_n denote the number of spanning trees in the fan graph shown in Figure 3.22. Verify that $h_1 = 1, h_2 = 3, h_3 = 8$.

Find a recurrence relation for h_n and hence show that $h_n = F_{2n-1}$.

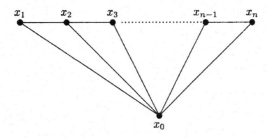

Figure 3.22

4
Travelling Round a Graph

In this chapter we consider various problems relating to the existence of certain types of walk in a graph. The reader should recall the definitions of walk, path, cycle and trail given in Section 3.2. The Königsberg bridge problem concerns the existence of a trail which is closed and contains all the edges of the graph. We study such (Eulerian) trails in more detail, but first we look at a related type of problem associated with the name of the Irish mathematician Sir William Rowan Hamilton (1805–1865).

4.1 Hamiltonian Graphs

The dodecahedron is shown at the end of Chapter 3. Hamilton posed the problem: is it possible to start at one of the 20 vertices, and, by following edges, visit every other vertex exactly once before returning to the starting point? In other words: is there a cycle through all the vertices? You should have no problem finding such a cycle (turn to Figure 4.3 if you get stuck), so it is perhaps not surprising that the commercial exploitation of this problem as a game was not a financial success.

Definition 4.1

A **hamiltonian cycle** in a graph G is a cycle containing all the vertices of G. A **hamiltonian graph** is a graph containing a hamiltonian cycle.

The name hamiltonian is, as often happens in mathematics, not entirely just, since others such as Kirkman had studied the idea before Hamilton.

Example 4.1

(a) The octahedral graph is hamiltonian: in Figure 4.1(a) take the hamiltonian cycle 1234561.

(b) The graph of Figure 4.1(b) is not hamiltonian. The easiest way to see this is to note that it has 9 vertices so that, if it is hamiltonian, it must contain a cycle of length 9. But, being a bipartite graph, it contains only cycles of even length.

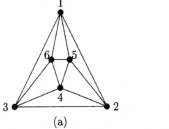 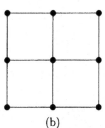

Figure 4.1

Theorem 4.1

A bipartite graph with an odd number of vertices cannot be hamiltonian.

Example 4.2

(a) K_n is hamiltonian for all $n \geq 3$.

(b) $K_{m,n}$ is hamiltonian if and only if $m = n \geq 2$.

(See Exercise 4.1.)

Example 4.3

The Petersen graph is not hamiltonian.

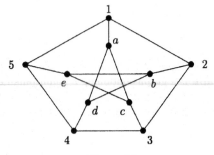

Figure 4.2

Solution

Label the vertices as shown in Figure 4.2, and suppose there is a hamiltonian cycle. Every time the cycle goes from the outside along one of the "spokes" $1a, 2b, 3c, 4d, 5e$, it has to return to the outside along another spoke. So the hamiltonian cycle must contain either 2 or 4 spokes.

(a) Suppose there are 4 spokes in a hamiltonian cycle: we can assume $5e$ is the one spoke not in it. Then 51 and 54 must be in the cycle, as must eb and ec. Since $1a$ and 15 are in the cycle, 12 is not, so 23 is. But this gives the cycle $2\,3\,c\,e\,b\,2$ as part of the hamiltonian cycle, which is clearly impossible.

(b) Suppose there are just two spokes in the hamiltonian cycle. Take $1a$ as one of them. Then ac or ad is in the cycle - say ad. Then ac is not, so $c3$ is. So spokes $b2, d4, e5$ are not in the cycle. Since $b2$ is not in the cycle, 23 must be. Similarly, since $d4$ is not in, 34 must be in the cycle. So all three edges from 3 are in the cycle, a contradiction.

There is no straightforward way of characterising hamiltonian graphs. Perhaps the best known simple sufficient condition is that given by Dirac in the following theorem, but it must be emphasised that the condition given is not at all necessary (as can be seen by considering the cycle C_n, $n \geq 5$).

Theorem 4.2 (Dirac, 1950)

If G is a simple graph with p vertices, each vertex having degree $\geq \frac{1}{2}p$, then G is hamiltonian.

Proof

Outlined in Exercise 4.6.

4.2 Planarity and Hamiltonian Graphs

There are some interesting connections between planar graphs and hamiltonian graphs. The first arose in connection with the Four Colour Conjecture (FCC), when it was realised that the presence of a hamiltonian cycle in a plane graph makes the colouring of its regions (faces) with four colours very easy. For example, consider the problem of colouring the faces of a dodecahedron using four colours. Figure 4.3 shows a hamiltonian cycle which divides the regions into an internal chain of regions, and an external chain. Colour the internal chain with colours A and B, and the external chain with colours C and D.

Early on in the history of the FCC, Tait conjectured that every polyhedral map in which every vertex has degree 3 has a hamiltonian cycle. (A map is polyhedral if any two adjacent regions meet in a single common edge or a

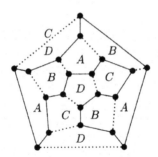

Figure 4.3 Dodecahedron

single point.) The truth of Tait's conjecture would have implied that every such map is 4-colourable; however, the conjecture was finally proved false in 1946, when Tutte constructed a counterexample.

Another connection between hamiltonicity and planarity occurs in the following algorithm which can be used to determine whether or not a given hamiltonian graph is planar. The basic idea is that if a graph G is both hamiltonian and planar, then, in a plane drawing of G, the edges of G which are not in the hamiltonian cycle H will fall into two sets, those drawn inside H and those drawn outside.

The planarity algorithm for hamiltonian graphs

1. Draw the graph G with a hamiltonian cycle H on the outside, i.e. with H as the boundary of the infinite region.

2. List the edges of G not in H: $e_1, \ldots, e_r$.

3. Form a new graph K in which the vertices are labelled $e_1, \ldots, e_r$ and where the vertices labelled e_i, e_j are joined by an edge if and only if e_i, e_j cross in the drawing of G, i.e. cannot both be drawn inside (or outside) H (such edges are said to be **incompatible**).

4. Then G is planar if and only if K is bipartite.

(If K is bipartite, with bipartition $B \cup W$, then the edges e_i coloured B can be drawn inside H, and the edges coloured W can be drawn outside.)

In practice, we introduce the edges one by one, as follows.

Example 4.4

Test the graph shown in Figure 4.4 for planarity.

Solution

1. The graph is already drawn with hamiltonian cycle $abcdefa$ on the outside.

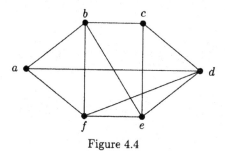

Figure 4.4

2. Edges not in the hamiltonian cycle are ad, be, bf, ce, df.

3. Start with ad; it is incompatible with bf, be, ce:

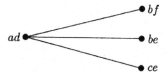

Now consider bf. It crosses only ad. Next consider be; it crosses df, so we get:

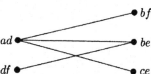

Now consider ce. It also crosses df, so we get:

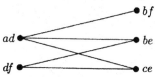

4. By now we have the full graph K. (Check: the number of edges in K is the number of crossings of edges in G.) Since K is bipartite, we conclude that G is planar, and we can draw it with ad and df inside, and bf, be, ce outside (Figure 4.5).

Example 4.5

Show that $K_{3,3}$ is not planar.

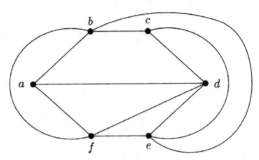

Figure 4.5

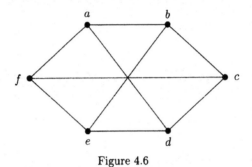

Figure 4.6

Solution

1. In Figure 4.6 we have $K_{3,3}$ drawn with hamiltonian cycle on the outside.

2. Edges not in hamiltonian cycle are ad, be, cf.

3. Obtain:

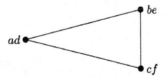

4. This is **not** bipartite, so $K_{3,3}$ is not planar.

4.3 The Travelling Salesman Problem

A sales representative of a publisher of mathematical texts has to make a round trip, starting at home, and visiting a number of university bookshops before returning home. How does the salesman choose his route to minimise the total distance travelled?

Here we consider a weighted graph, in which the vertices represent the book-shops and his home, and the edges represent the routes between them, each edge being labelled by the length of the route it represents. The salesman wishes to find a hamiltonian cycle of minimum length, i.e. of minimum total weight.

A complete graph K_n has $(n-1)!$ different hamiltonian cycles (or $\frac{1}{2}(n-1)!$ if we do not distinguish between a cycle and its "reverse"), so finding the one of minimum weight by looking at each in turn is out of the question when n is large. Even for $n = 10$, $\frac{1}{2}(n-1)! = 181\,440$. There is no really efficient algorithm yet known for solving the travelling salesman problem (TSP), so "good" rather than "best" routes are sought, as are estimates, rather than exact values, of the shortest total length.

Lower bounds

Lower bounds can be found by using spanning trees. First observe that if we take any hamiltonian cycle and remove one edge then we get a spanning tree, so

$$\text{Solution to TSP} > \text{minimum length of a spanning tree (MST).} \qquad (4.1)$$

But we can do better. Consider any vertex v in the graph G. Any hamiltonian cycle in G has to consist of two edges from v, say vu and vw, and a path from u to w in the graph $G - (v)$ obtained from G by removing v and its incident edges. Since this path is a spanning tree of $G - \{v\}$, we have

$$\text{Solution to TSP} \geq \left\{ \begin{array}{c} \text{sum of lengths of two} \\ \text{shortest edges from } v \end{array} \right\} + \left(\begin{array}{c} \text{MST of} \\ G - \{v\} \end{array} \right). \qquad (4.2)$$

Example 4.6

Apply (4.2) to the graph of Figure 4.7.

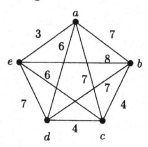

Figure 4.7

Solution

Choose vertex a. The two shortest edges from a have lengths 3 and 6. The minimum weight spanning tree of $G - \{a\}$ consists of edges bc, cd and ec, and has length 14. So, by (4.2), a lower bound for the TSP is $3 + 6 + 14 = 23$.

Instead, we could have started with b. The two shortest edges from b have lengths 4 and 7, and the minimum weight spanning tree of $G - \{b\}$ has length 13, so we obtain the lower bound $4 + 7 + 13 = 24$. This second bound gives us more information than the first.

Upper bounds

Assume that the weights are distances, satisfying the triangle inequality

$$d(x, z) \leq d(x, y) + d(y, z)$$

where $d(x, y)$ denotes the shortest distance along edges from x to y. In this case the following method gives upper bounds for the TSP in K_n.

Find a minimum spanning tree of K_n, say of weight w. We can then find a walk of length $2w$ which visits every vertex at least once, and which returns to its starting point, by going "round" the tree as shown in Figure 4.8.

Figure 4.8

We now try to reduce the length of this walk by taking shortcuts. Start at one vertex and follow the walk round. When we reach an edge which will take us to a vertex already visited, take the direct route to the next vertex not yet visited. For example, in Figure 4.8, which shows the minimum spanning tree of the graph of Example 4.6, we could start at a and obtain $aecbda$, which has length 26.

Since this method yields a hamiltonian cycle of length no greater than twice MST, we have

$$\text{MST} < \text{ solution to TSP} \leq 2\,\text{MST}, \qquad (4.3)$$

and, since MST < solution to TSP, by (4.1), we have constructed a hamiltonian cycle of length at most twice the minimum possible length. In Section 4.5 we shall improve this to at most $\frac{3}{2}$ times the minimum.

4.4 Gray Codes

A Gray code of order n is a cyclic arrangement of the 2^n binary sequences of length n such that any pair of adjacent sequences differ in only one place. For example, Figure 4.9(a) shows a Gray code of order 3.

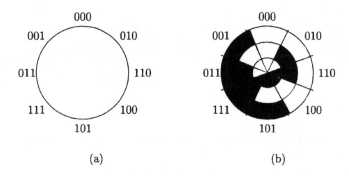

Figure 4.9

The industrial use of Gray codes is on account of their ability to describe the angular position of a rotating wheel. As in Figure 4.9(b), 0 and 1 are represented by white and black (off and on), and are read by electrical contact brushes. The fact that adjacent sequences differ in only one place reduces errors when the contact brushes are close to a boundary between segments. (Compare with 1999 changing to 2000 in a car milometer.)

Note that the code above corresponds to a hamiltonian cycle in a 3-dimensional cube (follow the arrows in Figure 4.10). Note also that the cycle involves going

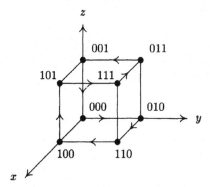

Figure 4.10

round the bottom of the cube (i.e. round a 2-dimensional cube!) with third coordinate 0, then moving up to change the third coordinate to 1, and then tracing out the 2-dimensional cube at the top, in the opposite direction. This idea generalises. So to obtain a Gray code of order 4, write down a Gray code of order 3 with 0 appended to each binary word, then follow it with the same Gray code of order 3, in reverse order, with 1 at the end of each word. This gives

$$0000 - 0100 - 1100 - 1000 - 1010 - 1110 - 0110 - 0010 - 0011 -$$
$$0111 - 1111 - 1011 - 1001 - 1101 - 0101 - 0001 - 0000.$$

4.5 Eulerian Graphs

The driver of a snow plough wishes to set out from the depot, travel along each road exactly once, and return to the depot. When is this possible? Similarly, the citizens of Königsberg wished to cross every bridge exactly once and return home. Both problems ask for a closed trail of a particular type.

Definition 4.2

An **eulerian circuit** is a closed trail which contains each edge of the graph. A graph which contains an eulerian trail is called an **eulerian graph**.

It was observed in Section 3.1 that a necessary condition for the existence of an eulerian circuit is that all vertex degrees must be even. It turns out that this condition is also sufficient in connected graphs. Our proof will use the following lemma.

Lemma 4.3

Let G be a graph in which every vertex has even degree. Then the edge set of G is an edge-disjoint union of cycles.

Proof

Proceed by induction on q, the number of edges. The lemma is true for $q = 2$, so consider a graph G with k edges and suppose that the lemma is true for all graphs with $q < k$. Take any vertex v_0, and start a walk from v_0, continuing until a vertex already visited is visited for the second time. If this vertex is v_j, then the part of the walk from v_j to v_j is a cycle C. Remove C to obtain a graph H with $< k$ edges and in which every vertex has even degree. By induction, H is an edge-disjoint union of cycles, so the result follows.

Theorem 4.4

Let G be a connected graph. Then G is eulerian if and only if every vertex has even degree.

Proof

⇒. Already shown.

⇐. Suppose every vertex has even degree. Then the edges fall into disjoint cycles. Take any such cycle C_1. If C_1 does not contain all the edges of G then, since G is connected, there must be a vertex $v_1 \in C_1$ and an edge $v_1 v_2$ not in C_1. Now $v_1 v_2$ is in some cycle, say C_2, disjoint from C_1. Insert C_2 into C_1 at v_1 to obtain a closed trail. If this trail does not contain all edges of G, take a vertex v_3 in $C_1 \cup C_2$ and edge $v_3 v_4$ not in $C_1 \cup C_2$. Then $v_3 v_4$ is in some cycle C_3 which we insert into $C_1 \cup C_2$. Continue in this way until all edges are used up.

Example 4.7

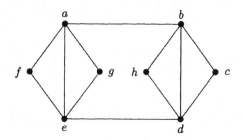

Figure 4.11

In Figure 4.11, first take cycle $abcdefa$. Then insert cycle $agea$ at a, and finally insert cycle $bdhb$ at b to obtain eulerian trail

$$a\,g\,e\,a\,b\,d\,h\,b\,c\,d\,e\,f\,a.$$

Definition 4.3

An **eulerian trail** is a trail which contains every edge of the graph, but is not closed. A non-eulerian graph which contains an eulerian trail is called a **semi-eulerian** graph.

The following result follows immediately from Theorem 4.4.

Theorem 4.5

A connected graph G is semi-eulerian if and only if it contains precisely two vertices of odd degree.

Example 4.8

In the Königsberg bridge problem, suppose that one further bridge is built. The resulting graph will then have two vertices of odd degree and hence will contain an eulerian trail.

An upper bound for the TSP

The following method yields a hamiltonian cycle in a complete graph whose length is at most $\frac{3}{2}$ times the length of the minimum hamiltonian cycle. This improves the bound in Section 4.3.

Given K_n, labelled by the length of the edges, first find a minimum spanning tree T. T must, by Exercise 3.1, have an **even** number $2m$ of vertices of odd degree. It is then possible to join these $2m$ vertices into m pairs by using m edges of K_n. Such a set of disjoint edges is called a **matching**. There will be many ways of choosing such a matching, so we choose a matching M of smallest total length. If we now add the edges of M to T, we obtain the new graph $M \cup T$ in which every vertex has even degree: thus $M \cup T$ possesses an eulerian circuit.

For example, with the graph of Example 4.7, T has length 17 (as in Figure 4.8) and T has four vertices of odd degree. Take $M = \{ad, bc\}$ to obtain $M \cup T$ as shown in Figure 4.12. An eulerian circuit is $aecbcda$.

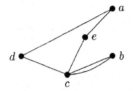

Figure 4.12

Starting at a, we can take $aecb$ and, to avoid visiting c twice, go directly from b to d, and then to a, obtaining the hamiltonian cycle $aecbda$ which has length 26.

We now show that the eulerian circuit obtained by this method always has length $\leq \frac{3}{2}$MST. Let TSP, EC, MST, M denote respectively the lengths of the minimum hamiltonian cycle, the eulerian circuit, the minimum spanning tree and the matching. Then

$$EC = MST + M, \quad TSP > MST.$$

The $2m$ vertices of M will occur, in some order, say $x_1, \dots, x_{2m}$, in the minimum length hamiltonian cycle. If for each $i < 2m$, we replace the part of the cycle between x_i and x_{i+1} by the edge $x_i x_{i+1}$, and we replace the part between x_{2m} and x_1 by the edge $x_{2m} x_1$, we obtain

$$\ell(x_1, x_2) + \ell(x_2, x_3) + \cdots + \ell(x_{2m}, x_1) \leq \text{TSP}$$

where $\ell(x_1, x_{i+1})$ denotes the length of the edge $x_i x_{i+1}$. Thus we have

$$(\ell(x_1, x_2) + \ell(x_3, x_4) + \cdots + \ell(x_{2m-1}, x_{2m})) + (\ell(x_2, x_3) + \cdots + \ell(x_{2m}, x_1)) \leq \text{TSP}.$$

So we obtain two matchings of $x_1, \ldots, x_{2m}$ whose lengths sum to $\leq$ TSP. One of these matchings must have total length $\leq \frac{1}{2}$ TSP, so that

$$M \leq \frac{1}{2} TSP.$$

Thus EC = MST + M $\leq$ TSP + $\frac{1}{2}$ TSP = $\frac{3}{2}$ TSP. Thus, on using shortcuts in the eulerian circuit to avoid repeating vertices, we obtain a hamiltonian cycle whose length is $\leq \frac{3}{2}$ TSP.

4.6 Eulerian Digraphs

A **digraph** or **directed graph** is a graph in which each edge is assigned a direction, indicated by an arrow. In place of the degree of a vertex we have the **indegree**, the number of edges directed towards the vertex, and the **outdegree**, the number of edges directed away from the vertex.

Example 4.9

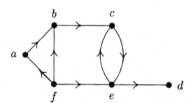

Figure 4.13

In Figure 4.13, the indegrees of $a, \ldots, f$ are respectively $1, 2, 2, 1, 2, 0$, and the outdegrees are $1, 1, 1, 0, 2, 3$. It should be clear why the sum of the indegrees equals the sum of the outdegrees.

An eulerian circuit in a digraph is exactly what we would expect; it has to follow the directions of the arrows at each stage. If every vertex has its indegree equal to its outdegree then, as in Lemma 4.3, the edge set can be partitioned as an edge-disjoint union of directed cycles, and, as in Theorem 4.4, we obtain:

Theorem 4.6

A connected digraph has an eulerian circuit if and only if each vertex has its indegree and outdegree equal.

Memory wheels

It is said that the meaningless Sanskrit word

yamátárájabhánasalagám

has been used as a memory aid by Indian drummers. It has in it every 3-tuple of
accented and unaccented vowels, each 3-tuple appearing once. We can display
this by replacing unaccented vowels by 0 and accented vowels by 1, to obtain

$$0\,1\,1\,1\,0\,1\,0\,0\,0\,1. \tag{4.4}$$

The 3-tuples $011, 111, 110, 101, 010, 100, 000, 001$ appear in it in this order. Note
that the last two digits of (4.4) are the same as the first two, so we can obtain
a "memory wheel" by overlapping the ends as shown in Figure 4.14.

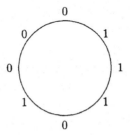

Figure 4.14

Now this arrangement achieves what a Gray code achieved, but much more
efficiently. A sensor placed at the edge of the wheel can read off triples of
digits and thereby determine how far the wheel has rotated. A Gray code for
8 positions would require three circles of 8 digits, i.e. 24 digits, whereas the
memory wheel uses only 8.

We now try to generalise this idea: can a circular arrangement of 2^n binary
digits be found which includes all 2^n n-digit binary sequences? One approach
might be via hamiltonian cycles. Since, in the above example, $1\underline{10}$ is followed
by $\underline{10}1$, and $1\underline{01}$ by $\underline{01}0$, we could take the triples xyz as the vertices of a graph
and join xyz and yzw by an edge to obtain the directed graph of Figure 4.15.

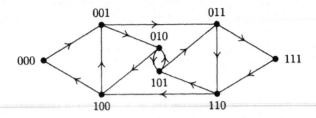

Figure 4.15

The directed hamiltonian cycle

$$000 - 001 - 011 - 111 - 110 - 101 - 010 - 100 - 000$$

yields the memory wheel of Figure 4.14. The trouble with this approach, however, is that it is not at all easy to see how to obtain a hamiltonian cycle in the
corresponding digraph when $n \geq 4$.

The problem was however solved by I.J. Good, in a 1946 paper in number
theory. Instead by taking the triples as the vertices, Good took the triples as
the edges of a graph, in which the vertices corresponded to the overlapping
2-tuples. So, for $n = 3$, we form the digraph of Figure 4.16.

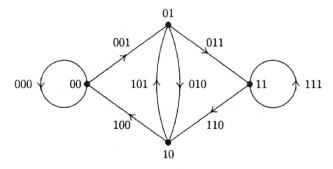

Figure 4.16

Now in this digraph all vertices have indegree and outdegree equal, so the
digraph contains an eulerian circuit. Once such circuit consists of the edges

$$000 - 001 - 011 - 111 - 110 - 101 - 010 - 100 - 000,$$

and this gives the same memory wheel as before.

In general, take as vertices the $(n-1)$-digit binary sequences, and draw a
directed edge from $x_1 x_2 \ldots x_{n-1}$ to $x_2 \ldots x_{n-1} x_n$, labelling the edge $x_1 x_2 \ldots x_n$.
The resulting digraph has an eulerian circuit which yields a memory wheel.

Example 4.10

Obtain a memory wheel containing all 16 4-digit binary sequences.

Solution

Construct a digraph with 8 vertices labelled by the eight 3-digit binary sequences, and draw a directed edge from $x_1 x_2 x_3$ to $x_2 x_3 0$ and to $x_2 x_3 1$. The
digraph of Figure 4.17 is obtained.

An eulerian circuit is (in terms of vertices)

$$000 - 000 - 001 - 011 - 111 - 111 - 110 - 101$$

$$-011 - 110 - 100 - 001 - 010 - 101 - 010 - 100 - 000$$

i.e. in terms of edges,

$$0000 - 0001 - 0011 - 0111 - 1111 - 1110 - 1101 - 1011$$

$$-0110 - 1100 - 1001 - 0010 - 0101 - 1010 - 0100 - 0000.$$

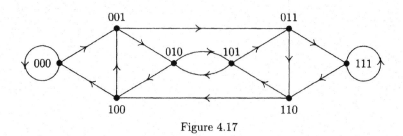

Figure 4.17

The corresponding memory wheel is as shown in Figure 4.18.

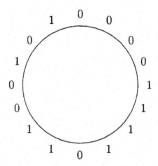

Figure 4.18

The problem of constructing memory wheels is also known as the rotating drum problem. The circular binary sequences are often called **maximum length shift register sequences**, or **de Bruijn sequences** after the Dutch mathematician N.G. de Bruijn who wrote about them in 1946 (although it turned out that they had been constructed many years before by C. Flye Sainte-Marie). They have been used worldwide in telecommunications, and there have been recent applications in biology.

Exercises

Exercise 4.1

(a) Strengthen Theorem 4.1 to: if a bipartite graph, with bipartition $V = B \cup W$, is hamiltonian, then $|B| = |W|$.

(b) Deduce that $K_{m,n}$ is hamiltonian if and only if $m = n \geq 2$.

Exercise 4.2

For each graph in Figure 4.19, determine whether (a) it is hamiltonian, (b) it is eulerian, (c) it is semi-eulerian.

(i)   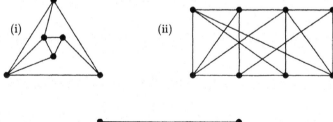   (ii)

(iii)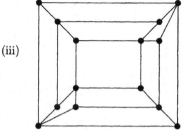

Figure 4.19

Exercise 4.3

Which of the platonic solid graphs are (a) hamiltonian, (b) eulerian?

Exercise 4.4

Use the planarity algorithm to determine whether or not the graphs in Figure 4.20 are planar.

Exercise 4.5

Construct a Gray code of order 5.

Exercise 4.6

Dirac's theorem. Prove Theorem 4.2 as follows. Suppose G is not hamiltonian. By adding edges we can assume that G is "almost" hamiltonian in the sense that the addition of any further edge will give a hamiltonian graph. So G has a path $v_1 \to v_2 \to \cdots \to v_p$ through every

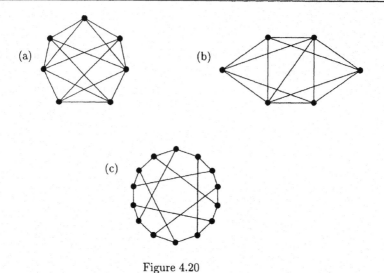

Figure 4.20

vertex, where v_1 and v_p are not adjacent. Show that there must be a vertex v_i adjacent to v_1, with v_{i-1} adjacent to v_n. This gives a hamiltonian cycle $v_1 \to \cdots \to v_{i-1} \to v_n \to \cdots \to v_{i+1} \to v_i \to v_1$.

Exercise 4.7

(a) **Ore's theorem.** Imitate the proof of Dirac's theorem to show that if G is a simple graph with $p \geq 3$ vertices, with $\deg(v) + \deg(w) \geq p$ for each pair of non-adjacent vertices v, w, then G is hamiltonian.

(b) Deduce that if G has $2 + \frac{1}{2}(p-1)(p-2)$ edges then G is hamiltonian.

(c) Find a non-hamiltonian graph with $1 + \frac{1}{2}(p-1)(p-2)$ edges.

Exercise 4.8

By removing vertex A, find a lower bound for the TSP for the graph of Exercise 3.7. Repeat, removing vertex B. Then obtain an upper bound by the method of Section 4.5.

Exercise 4.9

Find upper and lower bounds for the TSP for the situation in Exercise 3.8. How do your results compare with the exact solution?

Exercise 4.10

Construct a memory wheel containing all 32 5-digit binary sequences.

Exercise 4.11

Use digraphs to construct a memory wheel of length 9 containing all 2-digit ternary sequences (formed from the digits $0, 1, 2$). Then find one for all 3-digit ternary sequences.

Exercise 4.12

Dominoes. Can you arrange the 28 dominoes of an ordinary set in a closed loop, so that each matches with its neighbour in the usual way? Can you do so if all dominoes with a 6 on them are removed? Can you state a general theorem about dominoes with numbers $0, 1, \ldots, n$ on them? (Hint: consider each domino as an edge of a graph with vertices labelled $0, 1, \ldots, n$.)

Exercise 4.13

Figure 4.21 shows an arrangement of the numbers $1, \ldots, 5$ round a circle, so that each number is adjacent to every other number exactly once. Can you produce a similar arrangement for $1, \ldots, 7$? Use Euler's theorem to show that there is a solution for n numbers if and only if n is odd. Can you salvage a similar type of result when n is even?

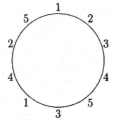

Figure 4.21

5

Partitions and Colourings

In this chapter we consider partitions of a set, introducing the Stirling numbers and the Bell numbers. We then consider vertex and edge colourings of a graph, where the vertex set and the edge set are partitioned by the colours.

5.1 Partitions of a Set

A **partition** of a set S is a collection of non-empty subsets $S_1, \ldots, S_r$ of S which are pairwise disjoint and whose union is S. The subsets S_i are called the **parts** of the partition. For example $\{1, 2, 4\} \cup \{3, 6\} \cup \{5\}$ is a partition of $\{1, \ldots, 6\}$ into three parts. Note that it does not matter in what order the parts appear.

Example 5.1

In a game of bridge, the 52 cards of a standard pack are distributed among four people who receive 13 cards each. In how many ways can the pack of 52 cards be partitioned into four sets of size 13?

Solution

We can choose 13 cards in $\binom{52}{13}$ ways. From the remaining 39 we can choose a further 13 in $\binom{39}{13}$ ways, and then from the remaining 26 we can choose 13 in $\binom{26}{13}$ ways. This leaves a final set of 13 cards. So we have

$$\binom{52}{13}\binom{39}{13}\binom{26}{13} = \frac{52!}{13!39!} \cdot \frac{39!}{13!26!} \cdot \frac{26!}{13!13!} = \frac{52!}{(13!)^4}$$

89

ways of partitioning the pack. But these partitions are not all distinct, since each distinct partition arises in 4! ways, depending on which of the four sets in it is chosen first, which is chosen second, and so on. So the required number is

$$\frac{52!}{(13!)^4 4!}$$

(a vast number, greater than 10^{27}).

There is another way of approaching this counting problem. Consider a row of 52 spaces grouped into four groups of 13:

$$(\ldots\ldots)\,(\ldots\ldots)\,(\ldots\ldots)\,(\ldots\ldots).$$

The cards can be placed in the spaces in 52! ways. Within each group there are 13! ways of arranging the same 13 cards, and these different arrangements are irrelevant since they give rise to the same part of the partition, so we have to divide by $(13!)^4$, one 13! for each group. Then the four groups themselves can be arranged in 4! ways, so we have to divide by 4!, giving the same answer as before.

This argument easily generalises, to give the following result.

Theorem 5.1

A set of mn objects can be partitioned into m sets of size n in

$$\frac{(mn)!}{(n!)^m m!}$$

different ways.

Corollary 5.2

A set of $2m$ objects can be partitioned into m pairs in

$$\frac{(2m)!}{2^m m!}$$

different ways.

Example 5.2

The number of ways of pairing 16 teams in a football cup draw is

$$\frac{16!}{2^8 8!} = 2\,027\,025.$$

The same type of argument can be applied when the parts of the required partition are not all of the same size.

Example 5.3

In how many ways can a class of 25 pupils be placed into four tutorial groups of size 3, two of size 4 and one of size 5?

Solution

Consider the following grouping of 25 spaces

$$(---)\,(---)\,(---)\,(---)\,(----)\,(----)\,(-----).$$

The 25 pupils can be placed in the spaces 25! ways. To count **distinct** partitions we have to take into account the ways of ordering the pupils within the groups - so we divide by $(3!)^4(4!)^25!$ - and also the ways of ordering the groups themselves - so we divide by 4! on account of the four groups of size 3 and by 2! on account of the two groups of size 4. So the required number is

$$\frac{25!}{(3!)^4(4!)^25!4!2!} \cong 3.6 \times 10^{15}.$$

Definition 5.1

A partition of an n-element set consisting of α_i subsets of size i, $1 \le i \le n$, where $\sum_{i=1}^{n} i\alpha_i = n$, is called a partition of **type** $1^{\alpha_1}2^{\alpha_2}\ldots n^{\alpha_n}$.

Generalising Example 5.3 gives the following result.

Theorem 5.3

The number of partitions of type $1^{\alpha_1}2^{\alpha_2}\ldots n^{\alpha_n}$ of an n-element set is

$$\frac{n!}{\prod_{i=1}^{n}(i!)^{\alpha_i}\alpha_i!}.$$

Example 5.4

The number of ways of grouping 10 people into two groups of size 3 and one group of size 4 is the number of partitions of type 3^24^1 and so is

$$\frac{10!}{(3!)^22!4!} = 2100.$$

5.2 Stirling Numbers

In this section we think about partitioning a set into a given number of parts.

Definition 5.2

Let $S(n,k)$ denote the number of ways of partitioning an n-set into exactly k parts. Then $S(n,k)$ is called a **Stirling number of the second kind**.

These numbers are named after the Scottish mathematician James Stirling (1692–1770), who is also known for his approximation of $n!$:

$$n! \sim \sqrt{2\pi n}\, n^n e^{-n}.$$

Stirling also has numbers of the first kind named after him – see Exercise 5.10. We now study $S(n, k)$. Clearly, for all $n \geq 1$,

$$S(n, 1) = S(n, n) = 1. \tag{5.1}$$

Example 5.5

We show that $S(4, 2) = 7$. Here are the seven ways of partitioning $\{1, 2, 3, 4\}$ into two parts: $\{1\} \cup \{2, 3, 4\}, \{2\} \cup \{1, 3, 4\}, \{3\} \cup \{1, 2, 4\}, \{4\} \cup \{1, 2, 3\}, \{1, 2\} \cup \{3, 4\}, \{1, 3\} \cup \{2, 4\}$ and $\{1, 4\} \cup \{2, 3\}$.

Clearly, for large n, we need a better way of evaluating $S(n, k)$ than just writing down all possible partitions. Such a method is given by the following recurrence relation.

Theorem 5.4

$$S(n, k) = S(n - 1, k - 1) + k\, S(n - 1, k) \tag{5.2}$$

whenever $1 < k < n$.

Proof

In any partition of $\{1, \ldots, n\}$ into k parts, the element n may appear by itself as a 1-element subset or it many occur in a larger set. If it appears by itself, then the remaining $n - 1$ elements have to form a partition of $\{1, \ldots, n - 1\}$ into $k - 1$ subsets, and there are $S(n - 1, k - 1)$ ways in which this can be done. On the other hand, if the element n is in a set of size at least two, we can think of partitioning $\{1, \ldots, n - 1\}$ into k sets - this can be done in $S(n - 1, k)$ ways - and then of introducing n into one of the k sets so formed - and there are k ways of doing this. So, by the addition and multiplication principles, we have $S(n, k) = S(n - 1, k - 1) + k\, S(n - 1, k)$.

Example 5.5 (again)

$$S(4, 2) = S(3, 1) + 2\, S(3, 2)$$

$$= 1 \quad + 2\,(S(2, 1) + 2\, S(2, 2))$$

$$= 1 \quad + 2(1 + 2) \quad = 7.$$

Theorem 5.5

For all $n \geq 2$, $S(n, 2) = 2^{n-1} - 1$.

Proof

We use induction on n. The result is true for $n = 2$, so suppose it is true for $n = k \geq 2$. Then

$$S(k + 1, 2) = S(k, 1) + 2\,S(k, 2) \quad \text{(by 5.2)}$$

$$= 1 + 2(2^{k-1} - 1)$$

$$= 1 + 2^k - 2 = 2^{(k+1)-1} - 1.$$

Table 5.1 gives the first few Stirling number $S(n, k)$.

Table 5.1

$n\backslash k$	1	2	3	4	5	6	7	8	$B(n)$
1	1								1
2	1	1							2
3	1	3	1						5
4	1	7	6	1					15
5	1	15	25	10	1				52
6	1	31	90	65	15	1			203
7	1	63	301	350	140	21	1		877
8	1	127	966	1701	1050	266	28	1	4140

Note the number $2^{n-1} - 1$ in the column $k = 2$. On the right of the table are the sums $B(n)$ of all the Stirling numbers in the rows. $B(n)$ is the total number of partitions of an n-set, and is called a **Bell number**, after another Scot, E.T. Bell, who emigrated to the USA. and wrote several popular books on mathematics, including *Men of Mathematics*, an idiosyncratic two-volume collection of "biographies" of famous mathematicians. We have, for $n \geq 1$,

$$B(n) = \sum_{k=1}^{n} S(n, k). \tag{5.3}$$

If we define $B(0) = 1 = S(0,0)$ (accept this as a useful convention, like $\binom{0}{0} = 1$), we can obtain a recurrence relation for the Bell numbers.

Theorem 5.6

For all $n \geq 1$, $B(n) = \sum_{k=0}^{n-1} \binom{n-1}{k} B(k)$.

The nth element of the set being partitioned will appear in one of the sets of the partition along with $j \geq 0$ other elements. There are $\binom{n-1}{j}$ ways of choosing these j elements. The remaining $n - 1 - j$ elements can then be partitioned in $B(n - 1 - j)$ ways. So

$$B(n) = \sum_{j=0}^{n-1} \binom{n-1}{j} B(n - 1 - j)$$

$$= \sum_{k=0}^{n-1} \binom{n-1}{k} B(k) \quad \text{(on putting } n - 1 - j = k\text{)}.$$

Example 5.6

$$B(9) = \sum_{k=0}^{8} \binom{8}{k} B(k)$$

$$= 1 + 8.1 + 28.2 + 56.5 + 70.15 + 56.52 + 28.203 + 8.877 + 1.4140$$

$$= 21\,147.$$

For an interesting (but useless!) formula for $B(n)$, see Exercise 5.9.

5.3 Counting Functions

The Stirling numbers arise naturally in the enumeration of all functions $f : X \to Y$ which can be defined from an m-set X to an n-set Y. There are n^m such functions since, for each $x \in X$, there are n possible values for $f(x)$.

Recall that the **image** of $f : X \to Y$ is the set of elements of Y which actually arise as a value $f(x)$ for some $x \in X$:

$$\text{im} f = \{y \in Y : y = f(x) \text{ for some } x \in X\}.$$

Each function $f : X \to Y$ has as its image a subset of Y. How many such functions have an image of size k? If f takes precisely k values then X can be partitioned into k parts, the ith of which will consist of those elements of X which are mapped onto the ith member of $\text{im}(f)$. So a function $f : X \to Y$ with image of size k can be constructed as follows:

(i) partition X into k parts $X_1, \ldots, X_k$ (this can be done in $S(m, k)$ ways);

(ii) choose the image set of size k in Y (this can be done in $\binom{n}{k}$ ways);

(iii) pair off each X_i with one of the members of the image set (this can be done in $k!$ ways).

So the number of functions $f : X \to Y$ with image of size k is $S(m, k)\binom{n}{k}k!$. Thus, since k can take any value from 1 to n, and since there are n^m functions $f : X \to Y$ altogether, we obtain:

Theorem 5.7

Let $|X| = m$ and $|Y| = n$ where $m, n \geq 1$.

(a) The number of functions $f : X \to Y$ with image of size k is $S(m, k)\binom{n}{k}k!$.
(b)

$$n^m = \sum_{k=1}^{n} S(m, k)\binom{n}{k}k!. \tag{5.4}$$

Note as a special case that the number of **surjections** from X to Y, i.e. functions whose image set is the whole of Y, is $n!S(m, n)$.

Example 5.7

We check (5.4) in the case $n = 4, m = 5$.

$$\sum_{k=1}^{4} S(5, k)\binom{4}{k}k! = 4S(5, 1) + 12S(5, 2) + 24S(5, 3) + 24S(5, 4)$$

$$= 4 + 180 + 600 + 240 = 1024 = 4^5.$$

Note that if we define $S(m, 0) = 0$ for all $m \geq 1$, and $S(0, 0) = 1$, then we can rewrite (5.4) as

$$n^m = \sum_{k=0}^{n} S(m, k)\binom{n}{k}k!.$$

This identity can be inverted.

Theorem 5.8

For all $m \geq 1, n \geq 0, m \geq n$,

$$n!S(m, n) = \sum_{k=0}^{n}(-1)^{n-k}\binom{n}{k}k^m. \tag{5.5}$$

Proof We can use Corollary 1.15, putting $a_k = k^m$ and $b_k = S(m,k)k!$.

Alternatively we shall be able to use the inclusion-exclusion principle in the next chapter: see Section 6.2.

Example 5.8

$$S(5,3) = \frac{1}{3!}\left(\sum_{k=0}^{3}(-1)^{3-k}\binom{3}{k}k^5\right) = \frac{1}{6}(-0 + 3 - 3.2^5 + 3^5) = 25.$$

5.4 Vertex Colourings of Graphs

To colour the vertices of a graph G is to assign a colour to each vertex in such a way that no two adjacent vertices receive the same colour. If we define an **independent** set of vertices of G to be a set of vertices no two of which are adjacent, then a vertex colouring can be thought of as a partition of the set V of vertices into independent subsets. Often we are concerned with the **smallest** number of colours required, i.e. the smallest number of independent sets which partition V; we call this number the chromatic number of G.

Definition 5.3

The **chromatic number** $\chi(G)$ of a graph G is the smallest value of k for which the vertex set of G can be partitioned into k independent subsets.

We have met the idea of colouring vertices already; in Section 3.5 we noted that bipartite graphs are bichromatic; so if G is bipartite with at least one edge then $\chi(G) = 2$. Also, the four colour theorem asserts that $\chi(G) \leq 4$ for all planar graphs G.

Theorem 5.9

(i) $\chi(K_n) = n$.

(ii) $\chi(C_n) = 2$ if n is even; $\chi(C_n) = 3$ if n is odd.

Proof

(i) No two vertices can receive the same colour since they are adjacent.

(ii) If n is even, we can alternate colours round the cycle; if n is odd we need a third colour for the "last" vertex coloured.

Example 5.9

The graph of Figure 5.1(a) has chromatic number 3; it needs at least three colours since it contains C_3, and three colours are sufficient, as shown in Figure 5.1(b).

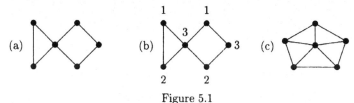

Figure 5.1

Note that the case of C_n, n odd, contradicts the belief of some amateur four-colour-theorem-provers, that a graph needs m colours only if it contains K_m as a subgraph. Another counterexample to this belief is the graph of Figure 5.1(c) which needs four colours (why?) although it does not contain K_4.

There is no easy way of finding $\chi(G)$ for a given graph G. The greedy algorithm, which we now describe, will give an upper bound for $\chi(G)$ related to the maximum vertex degree. In our description of the algorithm we denote colours by $C_1, C_2, C_3, \ldots$ and call C_i the ith colour.

The greedy algorithm for vertex colouring

1. List the vertices in some order: $v_1, \ldots, v_p$.

2. Assign colour C_1 to v_1.

3. At stage $i + 1$, when v_i has just been assigned a colour, assign to v_{i+1} the colour C_j with j as small as possible which has not yet been used to colour a vertex adjacent to v_{i+1}.

Example 5.10

We use the greedy algorithm to colour the graph of Figure 5.2 for each of the two vertex orderings shown.

With vertices listed as in (a), we assign colours as follows:

$$v: \quad 1 \quad 2 \quad 3 \quad 4 \quad 5 \quad 6 \quad 7$$

$$C: \quad 1 \quad 2 \quad 1 \quad 3 \quad 4 \quad 1 \quad 2$$

This colouring uses four colours. However, with the vertices labelled as in (b), we get:

$$v: \quad 1 \quad 2 \quad 3 \quad 4 \quad 5 \quad 6 \quad 7$$

$$C: \quad 1 \quad 2 \quad 3 \quad 1 \quad 3 \quad 1 \quad 2$$

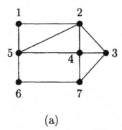   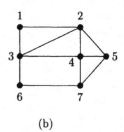

(a)   (b)

Figure 5.2

This second colouring shows that $\chi(G) \leq 3$; in fact $\chi(G) = 3$ since G is not bipartite.

Clearly, the bound for $\chi(G)$ obtained by the algorithm depends on the particular order in which the vertices are considered. But note that, if a vertex v has degree d then, when it is the turn of v to be assigned a colour, at most d of the colours are ineligible, so it must be given some colour C_i where $i \leq d + 1$. Thus we have the following bound.

Theorem 5.10

If G has maximum vertex degree Δ, then the greedy algorithm will colour the vertices of G using at most $\Delta + 1$ colours, so that $\chi(G) \leq \Delta + 1$.

Example 5.11 (A timetabling problem)

The University of Central Caledonia has nine vice-principals, Professors $A, B, \dots, I$, who serve on eight committees. The memberships of the committees are as follows.

Committee 1 :	A,	B,	C,	D	5 :	A,	H,	J	
2 :	A,	C,	D,	E	6 :	H,	I,	J	
3 :	B,	D,	F,	G	7 :	G,	H,	J	
4 :	C,	F,	G,	H	8 :	E,	I.		

Each committee is to meet for a day; no two committees with a member in common can meet on the same day. Find the smallest number of days in which the meetings can take place.

Solution

Represent each committee by a vertex, and join two vertices by an edge precisely when the corresponding committees have overlapping membership. Then the

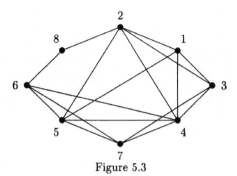

Figure 5.3

minimum number of days required is the chromatic number of the graph G, shown in Figure 5.3. Note that vertices $1, 2, 3, 4$ form a K_4, so at least four colours (days) are needed. But four colours are sufficient: e.g.

$$\{1, 7, 8\} \cup \{3, 5\} \cup \{2, 6\} \cup \{4\}$$

is a partition of $\{1, \ldots, 8\}$ into independent sets. So $\chi(G) = 4$, and four days are enough.

5.5 Edge Colourings of Graphs

An **edge colouring** of a graph G is an assignment of colours to the edges of G so that no two edges with a common vertex receive the same colour. The minimum number of colours required in an edge colouring of G is called the **chromatic index** of G and is denoted by $\chi'(G)$.

Thus to edge colour a graph is to partition the edge set into subsets such that no two edges in the same subset have a vertex in common, i.e. so that all edges in any part of the partition are **disjoint**. A set of disjoint edges in a graph is often called a **matching**. Clearly, in an edge colouring, all edges at a vertex v must receive different colours, so $\chi'(K_n) \geq n - 1$ for each n.

Example 5.12

(a) $\chi'(K_4) = 3$, since $\chi'(K_4) \geq 3$ and three colours suffice, as shown in Figure 5.4(a).
(b) $\chi'(K_5) = 5$. Here, $\Delta = 4$ colours are not enough. For there are 10 edges and no more than two edges in any matching. However, 5 colours are enough, as shown in Figure 5.4(b).

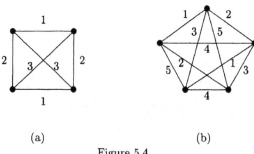

(a)   (b)

Figure 5.4

Theorem 5.11

(i) $\chi'(K_n) = n$ if n is odd.

(ii) $\chi'(K_n) = n - 1$ if n is even.

Proof

(i) If n is odd, any matching in K_n can have at most $\frac{1}{2}(n-1)$ edges. So at most $\frac{1}{2}(n-1)$ edges can be given any one colour. But there are $\frac{1}{2}n(n-1)$ edges in K_n, so at least n colours are needed. Now we can colour the edges using n colours in the following way. Represent K_n as a regular n-gon, with all diagonals drawn. Colour the boundary edges by $1, \ldots, n$; then colour each diagonal by the colour of the boundary edge parallel with it. This gives an edge colouring using n colours. The case $n = 5$ is as in Figure 5.4(b).

(ii) Now suppose n is even. Certainly $\chi'(K_n) \geq n - 1$; we show how to use only $n - 1$ colours. Since $n - 1$ is odd, we can colour K_{n-1} using $n - 1$ colours, as described above. Now take another vertex v and join each vertex of K_{n-1} to v, thus obtaining K_n. At each vertex of K_{n-1}, one colour has not been used. The colours missing at each vertex of K_{n-1} are all different, so we can use these $n - 1$ colours to colour the added edges at v. This gives an edge colouring of K_n using $n - 1$ colours.

The appearance of $\Delta(= n - 1)$ and $\Delta + 1(= n)$ as the chromatic indices of K_n, according as n is even or odd, is in accordance with the following result.

Theorem 5.12 (Vizing, 1964)

If G is a simple graph with maximum vertex degree Δ, then $\chi'(G) = \Delta$ or $\Delta + 1$.

We omit the proof of this result; a proof can be found in [9]. But we include the statement of the result because it has led to a great deal of work on determining which graphs are **class one** graphs, i.e. satisfy $\chi'(G) = \Delta$, and which are **class two**, i.e. satisfy $\chi'(G) = \Delta + 1$.

Example 5.13

The Petersen graph is class 2. Here $\Delta = 3$, so we have to show that $\chi'(G) \neq 3$. So suppose an edge colouring using only three colours exists. Then the outer 5-cycle uses three colours, and, without loss of generality, we can assume that it is coloured as in Figure 5.5(a). The spokes are then uniquely coloured, as in

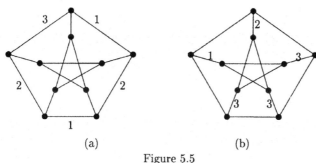

(a)   (b)

Figure 5.5

Figure 5.5(b). But this leaves two adjacent inside edges which have to be given colour 2. So there is no edge colouring with three colours.

We close this section by establishing that all bipartite graphs are class 1. This result is due to König, the Hungarian author of the first major book on graph theory [14].

Theorem 5.13 (König)

$\chi'(G) = \Delta$ for all bipartite graphs G.

Proof

Proceed by induction on q, the number of edges. The theorem is clearly true for graphs with $q = 1$; so suppose it is true for all bipartite graphs with k edges, and consider a bipartite graph G with maximum vertex degree Δ and with $k + 1$ edges. Choose any edge vw of G, and remove it, thereby forming a new bipartite graph H. H has k edges and maximum vertex degree $\leq \Delta$, so, by the induction hypothesis, H can be edge coloured using at most Δ colours.

Now, in H, v and w both have degree $\leq \Delta - 1$, so there is at least one colour missing from the edges from v, and at least one missing from the edges of w. If there is a colour missing at both vertices then it can be used to colour edge vw. If there is no colour missing from both, then let C_1 be a colour missing at v, and C_2 a colour missing at w. Now there is some edge, say vu, coloured C_2; if there is an edge coloured C_1 from u, go along it, and continue along edges coloured C_1 and C_2 alternately as far as possible. The path so constructed will never reach w since if it did it would have to reach w along an edge coloured C_1 and so would be a path of even length, giving, with edge vw, an odd cycle in a bipartite graph. So the connected subgraph K, consisting of vertex v and all vertices and edges of H which can be reached by a path of edges coloured C_1 and C_2, does not contain w. So we can interchange the colours C_1 and C_2 in K without interfering with the colours in the rest of H. This gives a new edge colouring of H in which v and w have no edge coloured C_2, and we can use C_2 to colour vw.

This idea of swapping colours along a path was used by Kempe in his unsuccessful 1879 attempt to prove the four colour theorem. Despite the fact that it did not work there as Kempe had hoped, it nevertheless has proved to be a very useful technique in graph theory.

Example 5.14

Eight students require to consult certain library books. Each is to borrow each required book for a week. The books B_j required by each student S_i are as follows:

$$S_1 : B_1, B_2, B_3 \qquad S_2 : B_2, B_4, B_5, B_6 \qquad S_3 : B_2, B_3, B_5, B_7$$

$$S_4 : B_3, B_5 \qquad S_5 : B_1, B_6, B_7 \qquad S_6 : B_2, B_4, B_6$$

$$S_7 : B_4, B_5, B_7 \qquad S_8 : B_3, B_6.$$

What is the minimum number of weeks required so that each student can borrow all books required?

Solution

Draw a bipartite graph G with vertices labelled $B_1, \ldots, B_7, S_1, \ldots, S_8$, and with S_i joined by an edge to B_j precisely when student S_i has to consult book B_j. Then G has maximum vertex degree $\Delta = 4$, so, by König's theorem, $\chi'(G) = 4$. Thus four colours (weeks) are required. You should be able to partition the set of edges into four disjoint matchings.

Exercises

Exercise 5.1

How many ways are there of arranging 16 football teams into four groups of four?

Exercise 5.2

A class contains 30 pupils. For a chemistry project, the class is to be put into four groups, two of size 7 and two of size 8. In how many ways can this be done?

Exercise 5.3

In the early versions of the Enigma machine, used in Germany in the 1930s, the plugboard swapped six pairs of distinct letters of the alphabet. In how many ways can this be done (assuming 26 letters)?

Exercise 5.4

Any permutation is a product of cycles. For example, the permutation 351642 $(3 \rightarrow 1, 5 \rightarrow 2, 1 \rightarrow 3, 6 \rightarrow 4, 4 \rightarrow 5, 2 \rightarrow 6)$ can be written as $(31)(2645)$. How many permutations of $1, \ldots, 8$ are a product of a 1-cycle, two 2-cycles and a 3-cycle?

Exercise 5.5

Prove that (a) $S(n, n-1) = \binom{n}{2}$, (b) $S(n, n-2) = \binom{n}{3} + 3\binom{n}{4}$.

Exercise 5.6

Prove by induction that $S(n, 3) > 3^{n-2}$ for all $n \geq 6$.

Exercise 5.7

Show that $S(n, k) = \sum_{m=k-1}^{n-1} \binom{n-1}{m} S(m, k-1)$ and hence given another proof of Theorem 5.6.

Exercise 5.8

Find $B(10)$.

Exercise 5.9

Use Theorem 5.6 and induction to prove that $B(n) = \frac{1}{e} \sum_{j=0}^{\infty} \frac{j^n}{j!}$.

Exercise 5.10

The (signless) Stirling numbers $s(n, k)$ of the first kind are defined by: $s(n, k)$ is the number of permutations of $1, \dots, n$ consisting of exactly k cycles. Verify that $s(2, 1) = 1, s(3, 1) = 2, s(3, 2) = 3, s(4, 2) = 11$ and that $s(n, 1) = (n - 1)!$. Prove that $s(n, k) = (n - 1)s(n - 1, k) + s(n - 1, k - 1)$, and deduce the value of $s(6, 2)$.

Exercise 5.11

Find $\chi(G)$ and $\chi'(G)$ for each of the graphs of Exercise 4.2.

Exercise 5.12

Let G be a graph with p vertices and let $\alpha(G)$ denote the size of the largest independent set of vertices of G. Show that $\chi(G)\alpha(G) \geq p$.

Exercise 5.13

Apply the greedy vertex colouring algorithm to the graph of Figure 5.3, taking the vertices (a) in the order $1, \dots, 8$, (b) in order $8, \dots, 1$. Do you get a colouring using four colours?

Exercise 5.14

As Exercise 5.13, but this time choose vertices in (a) increasing, (b) decreasing order of vertex degrees. Which approach would you expect to require fewer colours in general?

Exercise 5.15

Explain why there is always an ordering of the vertices for which the greedy algorithm will lead to a colouring with $\chi(G)$ colours.

Exercise 5.16

Find the chromatic index of each of the five Platonic solid graphs.

Exercise 5.17

A graph in which every vertex degree is 3 is called a **cubic** graph. Prove that all hamiltonian cubic graphs have chromatic index 3. (Note however that not all cubic graphs have chromatic index 3, e.g. the Petersen graph.)

Exercise 5.18

Let G be a graph with an odd number $p = 2k + 1$ of vertices, each of which has the same degree r.
(a) Show that G has $(k + \frac{1}{2})r$ edges.
(b) Explain why no more than k edges can have the same colour in any edge colouring, and hence show that $\chi'(G) = r + 1$. Thus every regular graph with an odd number of vertices is class 2. (This includes K_n, n odd, as shown in Theorem 5.11.)

Exercise 5.19

Let $f_\lambda(G)$ denote the number of ways of colouring the vertices of G using λ given colours.
(a) Show that $f_\lambda(K_n) = \lambda(\lambda - 1)(\lambda - 2)\ldots(\lambda - n + 1)$.
(b) Show that $f_\lambda(T) = \lambda(\lambda - 1)^{n-1}$ for all trees T with n vertices.
(c) Let xy be any edge of G. Let G' be the graph obtained from G by removing the edge xy, and let G'' be the graph obtained by identifying vertices x and y. Then $f_\lambda(G) = f_\lambda(G') - f_\lambda(G'')$. Deduce that $f_\lambda(G)$ is a polynomial in λ: it is called the **chromatic polynomial** of G.
(d) Note that the solution $a_n = 2^n + (-1)^n 2$ of Example 2.4 can be interpreted as: $f_3(C_n) = 2^n + (-1)^n 2$. By replacing 3 colours by λ colours, show similarly that $f_\lambda(C_n) = (\lambda - 1)^n + (-1)^n(\lambda - 1)$. Note that this gives $f_2(C_n) = 0$ whenever n is odd, as expected!

6

The Inclusion–Exclusion Principle

In this chapter we discuss a method of counting which has been used for at least 300 years. One of its first uses was in the study of derangements; as well as this, we give many applications including labelled trees, scrabble and the ménage problem.

6.1 The Principle

The principle is essentially a generalisation of the following simple observation. Suppose we are given two sets A and B, and are asked for the number of elements in their union. A first attempt might be $|A| + |B|$, but elements in both A and B are counted twice; so the corrected estimate is

$$|A \cup B| = |A| + |B| - |A \cap B|. \qquad (6.1)$$

Note that we first **include**, and then **exclude** all those which have been included too often.

Even this simplest form of the principle can be useful.

Example 6.1

In a class of 50, there are 30 girls, and there are 35 students with dark hair. Show that there are at least 15 girls with dark hair.

Solution

Let A denote the set of female students, and B the set of students with dark hair. Then

$$|A \cap B| = |A| + |B| - |A \cup B| = 30 + 35 - |A \cup B|$$
$$\geq 65 - 50 = 15 \qquad \text{since } |A \cup B| \leq 50.$$

The next application of (6.1) is much less trivial.

Example 6.2

We look for the smallest possible value of m such that if G is a graph on 60 vertices, each vertex having degree at least m, then G must contain K_4 as a subgraph.

It is not hard to see that m must be greater than 40. For if we take G to be the graph $K_{20,20,20}$, with vertex set $V = V_1 \cup V_2 \cup V_3$, $|V_i| = 20$ for each i, all vertices of V_i being joined to all vertices of V_j whenever $i \neq j$, then each vertex will have degree 40; but no K_4 exists in G since any K_4 in G would have to have two vertices in the same V_i. We now show that $m = 41$. Let G be any graph on 60 vertices, each of degree > 40. Choose any vertex v_1, and let S_1 be the set of vertices G adjacent to v_1. Then $|S_1| > 40$. Take any vertex v_2 in S_1, and let S_2 be the set of all vertices adjacent to v_2. Then

$$|S_1 \cap S_2| = |S_1| + |S_2| - |S_1 \cup S_2|$$
$$> 40 + 40 - |S_1 \cup S_2| \geq 80 - 60 = 20.$$

So $|S_1 \cap S_2| > 20$. Next take any $v_3 \in S_1 \cap S_2$, and let S_3 be the set of all vertices of G adjacent to v_3. Then

$$|S_1 \cap S_2 \cap S_3| = |(S_1 \cap S_2) \cap S_3| = |S_1 \cap S_2| + |S_3| - |(S_1 \cap S_2) \cup S_3|$$
$$> 20 + 40 - 60 = 0.$$

Thus there exists a vertex v_4 in $S_1 \cap S_2 \cap S_3$. But then v_1, v_2, v_3, v_4 are all adjacent to each other in G, so that G contains a K_4.

We now extend (6.1) to three sets A, B, C as shown in Figure 6.1. Our first estimate for $|A \cup B \cup C|$ might be $|A| + |B| + |C|$. But elements in more than one set will have been included more than once, so a second estimate might be $|A| + |B| + |C| - |A \cap B| - |B \cap C| - |C \cap A|$. But elements in all three sets A, B, C will have then been included thrice and excluded thrice, and so have to be included once more. So finally we get

$$|A \cup B \cup C| = |A| + |B| + |C| - |A \cap B| - |B \cap C| - |C \cap A| + |A \cap B \cap C|. \quad (6.2)$$

Note that we **include**, then **exclude**, then **include** again.

The general formulation is as follows.

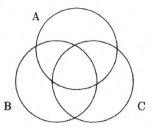

Figure 6.1

Theorem 6.1 (The inclusion-exclusion principle)

Let S be a set of objects, and let $P_1, \ldots, P_r$ be properties which the elements of S may or may not possess. Let $N(i, j, \ldots, k)$ denote the number of elements of S which possess properties $P_i, P_j, \ldots, P_k$ (and possibly some others as well). Then the number of elements of S possessing at least one of the properties is

$$\sum_i N(i) - \sum_{i<j} N(i,j) + \sum_{i<j<k} N(i,j,k) - \ldots + (-1)^{r-1} N(1, 2, \ldots, r). \quad (6.3)$$

Proof

Any element of S which has none of the properties contributes 0 to each term of (6.3) and hence contributes 0 to the sum.

Now consider an element of S which possesses $t \geq 1$ of the properties, where $t \leq r$. It contributes to t of the terms $N(i)$, to $\binom{t}{2}$ of the terms $N(i,j)$, and so on. So its total contribution to (6.3) is

$$t - \binom{t}{2} + \binom{t}{3} - \cdots + (-1)^{t-1} \binom{t}{t} = 1 - \{1 - t + \binom{t}{2} - \cdots + (-1)^t \binom{t}{t}\}$$

$$= 1 - (1 - 1)^t = 1.$$

Example 6.3

Find the number of positive integers ≤ 100 which are divisible by 3 or 7.

Solution

Take $S = \{1, \ldots, 100\}$, let P_1 be the property of being divisible by 3 and let P_2 be the property of being divisible by 7. Then $N(1)$ is the number of integers ≤ 100 which are divisible by 3, so $N(1) = 33$. Similarly $N(2) = 14$. Finally $N(1,2)$ is the number of such integers divisible by both 3 and 7, i.e. divisible by 21; so $N(1,2) = 4$. Thus the required number is $33 + 14 - 4 = 43$.

The principle is in fact more often used in a slightly different form. Instead of asking how many elements have at **least one** of the properties, we ask how many have **none** of the properties.

Theorem 6.2 (Second version of the inclusion-exclusion principle)

With the same notation as in Theorem 6.1, the number of elements of S possessing **none** of the properties is

$$|S| - \sum_i N(i) + \sum_{i<j} N(i,j) - \ldots + (-1)^r N(1,2,\ldots,r). \qquad (6.4)$$

Example 6.4

How many positive integers ≤ 100 are divisible by none of $2, 3, 5, 7$?

Solution

Here again $S = \{1, \ldots, 100\}$ and $P_1, \ldots, P_4$ are the properties of being divisible by $2, 3, 5, 7$ respectively. Then $N(1) = 50, N(2) = 33, N(3) = 20$ and $N(4) = 14$. Next, $N(1,2)$ is the number of elements of S divisible by 2 and 3, i.e. by 6, so $N(1,2) = 16$. Similarly, for example, $N(1,3,4)$ is the number of positive integers ≤ 100 divisible by $2, 5$ and 7, i.e. by 70, so that $N(1,3,4) = 1$. Thus the required number is

$$100-(50+33+20+14)+(16+10+7+6+4+2)-(3+2+1) = 100-117+45-6 = 22.$$

What is the significance of this result? Observe that any number ≤ 100 which is not prime must possess a factor $\leq \sqrt{100} = 10$, and so must be divisible by a prime ≤ 10, i.e. by $2, 3, 5$ or 7. The answer 22 just obtained is not quite the number of primes ≤ 100 since it does not include the primes $2, 3, 5, 7$ themselves, and it also includes 1 which is not prime. So the number of primes ≤ 100 is $22 + 4 - 1 = 25$. Check it!

Example 6.5

In one version of the Enigma machine used for encoding secret messages, three "rotors" were chosen from a set of five, and were placed in order in the machine. On each day a different ordered set of three was chosen, such that no rotor was in the same position as on the previous day. Given the arrangement for one day, how many possible ordered choices are there for the next day?

Solution

Let us suppose that the rotors are labelled $1, \ldots, 5$, and that on a given day rotors $1, 2, 3$ are chosen, in that order. Let S denote the set of all possible ordered sets of three rotors; then $|S| = 5 \times 4 \times 3 = 60$. For each $i \leq 3$, let

P_i denote the property that rotor i is in position i. Then we want the number of members of S possessing none of the properties P_1, P_2, P_3. So the required number is

$$|S| - \sum N(i) + \sum N(i,j) - N(1,2,3).$$

Now $N(i) = 4 \times 3$ for each $i \le 3$, and each $N(i,j) = 3$; so the answer is

$$60 - 3 \times (4 \times 3) + 3 \times 3 - 1 = 32.$$

(Later on the number of rotors used by the German navy was increased to 8; see Exercise 6.9.)

The next example deals with derangements, already met in Chapter 2.

Example 6.6

Recall that a **derangement** of n objects is a permutation of them with the property that no object is in its original place. We now show how to use the inclusion–exclusion principle to derive the formula (2.10) for d_n, the number of derangements of n objects.

Take S to be the set of all permutations of $1, \ldots, n$ and, for each $i \le n$, let P_i be the property: i is in the ith position. Then d_n is the number of members of S possessing none of the properties P_i. Now, for each i, $N(i) = (n-1)!$ since i is fixed and the remaining $n-1$ numbers can be permuted in any way we like. Similarly, $N(i,j) = (n-2)!$, and so on. Note also that there are $\binom{n}{1}$ terms $N(i)$, $\binom{n}{2}$ terms $N(i,j)$, etc., so that, by (6.4),

$$d_n = |S| - \sum_i N(i) + \sum_{i<j} N(i,j) - \cdots + (-1)^n N(1, \ldots, n)$$

$$= n! - \binom{n}{1}(n-1)! + \binom{n}{2}(n-2)! - \cdots + (-1)^n \binom{n}{n} 0!$$

$$= n! - \frac{n!}{1!} + \frac{n!}{2!} - \cdots + (-1)^n \frac{n!}{n!}$$

$$= n!\{1 - \frac{1}{1!} + \frac{1}{2!} - \cdots + \frac{(-1)^n}{n!}\}.$$

The next example introduces ideas which will be useful in section 6.4.

Example 6.7

How many non-negative integer solutions are there of the equation $x+y+z = 20$ satisfying the conditions $x \le 10, y \le 5, z \le 15$?

Solution

Let S denote the set of all non-negative solutions of $x+y+z = 20$. By Theorem

1.11, $|S| = \binom{22}{20} = \binom{22}{2}$. Let P_1 be the property $x \geq 11$; let P_2 be the property $y \geq 6$, and P_3 the property $z \geq 16$. Then we want the number of members of S possessing **none** of the properties P_i. This number is

$$\binom{22}{2} - \sum_i N(i) + \sum_{i<j} N(i,j) - N(1,2,3). \tag{6.5}$$

Consider $N(1)$. If $x \geq 11$, $x = 11 + u$ for some $u \geq 0$, and the equation becomes $u + y + z = 9$. The number of non-negative integer solutions of this equation is $N(1) = \binom{11}{9} = \binom{11}{2}$. Similarly, $N(2) = \binom{16}{2}, N(3) = \binom{6}{2}$. Next, $N(1,2)$ is the number of solutions satisfying $x \geq 11$ and $y \geq 6$; putting $x = 11 + u, y = 6 + v$ changes the equation into $u + v + z = 3$, and the number of non negative solutions is $N(1,2) = \binom{5}{2}$. Since $N(1,3) = N(2,3) = N(1,2,3)$ are all clearly zero, we find that (6.5) becomes

$$\binom{22}{2} - \binom{11}{2} - \binom{16}{2} - \binom{6}{2} + \binom{5}{2} = 51.$$

6.2 Counting Surjections

Let $|X| = m$ and $|Y| = n$, and consider functions $f : X \to Y$. The image of f was defined in Section 5.3 to be the set of all elements of Y which arise as $f(x)$ for some $x \in X$. We say that f is **surjective** if the image of f is the whole of Y. How many surjections $f : X \to Y$ are there?

Let S denote the set of all functions $f : X \to Y$, where $Y = \{y_1, \dots, y_n\}$, and let P_i be the property: y_i is not in the image of f. Then $N(1) = (n-1)^m$, since each of the m elements of X can be mapped by f onto any of the $n-1$ other elements of Y. Similarly, $N(i_1, \dots, i_k) = (n-k)^m$. So by (6.4), the number of surjections from X to Y, i.e. the number of members of S possessing **none** of the properties P_i is

$$|S| - \sum_i N(i) + \sum_{i<j} N(i,j) - \cdots + (-1)^n N(1,\dots,n)$$

$$= n^m - \binom{n}{1}(n-1)^m + \binom{n}{2}(n-2)^m - \cdots + (-1)^n \binom{n}{n}(n-n)^m$$

$$= \sum_{i=0}^{n-1} (-1)^i \binom{n}{i}(n-i)^m.$$

One immediate consequence of this result arises from the observation that, if $n > m$, there are **no** surjections $f : X \to Y$ since there are not enough elements in X to be mapped onto the n elements of Y. So we obtain:

$$\sum_{i=0}^{n-1} (-1)^i \binom{n}{i}(n-i)^m = 0 \qquad \text{whenever } m < n. \tag{6.6}$$

Another consequence arises from the fact that the number of surjections from X to Y is $n!S(m,n)$ (as in Theorem 5.7); so we obtain another proof of (5.5).

6.3 Counting Labelled Trees

It was mentioned in Section 3.3 that Cayley proved that the number of labelled trees on n vertices is n^{n-2}. The three trees on 3 points were shown in Figure 3.7. Many proofs of Cayley's result have been given. Most of these proofs are quite tricky; the most common one, due to Prüfer, depends upon constructing a one-to-one correspondence between the trees and ordered $(n-2)$-tuples of numbers each of which can be any of $1, \ldots, n$. However, we are here going to use ideas of J. Moon and employ the inclusion-exclusion principle and (6.6).

Let S denote the set of all spanning trees on the vertices labelled $1, \ldots, n$, and let $|S| = T(n)$. For each $i \le n$, let P_i be the property: vertex i is an end vertex. Since, by Theorem 3.3, every tree with $p \ge 2$ vertices has an end vertex, every member of S must possess at least one of the properties P_i. Further, if $n \ge 3$, no tree on n vertices can have every vertex an end vertex, so no member of S can possess all n properties. So, by (6.3), for $n \ge 3$,

$$T(n) = \sum_i N(i) - \sum_{i<j} N(i,j) + \cdots + (-1)^n \sum_{i_1 < \ldots < i_{n-1}} N(i_1, \ldots, i_{n-1}).$$

Now $N(i) = (n-1)T(n-1)$, since if vertex i is an end vertex, its edge can go to any of the other $n-1$ vertices, and these $n-1$ vertices are joined by a tree.

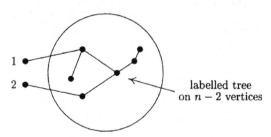

1

2   labelled tree
                                                       on $n-2$ vertices

Figure 6.2 $N(1,2) = (n-2)^2 T(n-2)$

Similarly, as illustrated in Figure 6.2, $N(i,j) = (n-2)^2 T(n-2)$, and so on. So, for $n \ge 3$,

$$T(n) = \binom{n}{1}(n-1)T(n-1) - \binom{n}{2}(n-2)^2 T(n-2)$$

$$+ \cdots + (-1)^n \binom{n}{n-1} T(1)$$

$$= \sum_{i=1}^{n-1}(-1)^{i-1}\binom{n}{i}(n-i)^i T(n-i). \tag{6.7}$$

However, putting $m = n - 2$ in (6.6) and rearranging slightly, we get

$$n^{n-2} = \sum_{i=1}^{n-1}(-1)^{i-1}\binom{n}{i}(n-i)^i(n-i)^{n-i-2}. \tag{6.8}$$

Compare (6.7) and (6.8). If the formula $T(k) = k^{k-2}$ is known to be true for all k up to $n - 1$, then the right-hand sides of (6.7) and (6.8) coincide, and it therefore follows that $T(n) = n^{n-2}$. Thus, since the result is true for $n = 3$, it follows by induction that $T(n) = n^{n-2}$ for all $n \geq 3$.

6.4 Scrabble

Scrabble is a word game in which players take turns to use the letters in their possession to form new words. At each stage of the game, each player has 7 tiles, each tile having on it a letter or being blank. The distribution of letters is as follows.

A	B	C	D	E	F	G	H	I	J	K	L	M
9	2	2	4	12	2	3	2	9	1	1	4	2

N	O	P	Q	R	S	T	U	V	W	X	Y	Z	blank
6	8	2	1	6	4	6	4	2	2	1	2	1	2

At the start of the game, each player chooses 7 tiles. How many ways are there of choosing 7 tiles?

If we let a denote the number of As, b the number of Bs, ..., z the number of Zs, and ω the number of blanks, then the number of possible choices of 7 tiles is just the number of solutions of the equation

$$a + b + \ldots + z + \omega = 7 \tag{6.9}$$

in non-negative integers, with $a \leq 9, b \leq 2$, and so on. So, to apply the inclusion-exclusion principle, take S to be the set of all non-negative integer solutions of (6.9), take P_a to be the property that $a \geq 10, P_b$ to be the property that $b \geq 3$, and so on: then we want the number of members of S with none of the properties, as given by (6.4). We proceed as in Example 6.7.

By Theorem 1.11 , $|S| = \binom{27+7-1}{7} = \binom{33}{7}$.

Next consider $N(a)$, the number of solutions of (6.9) with $a \geq 10$. This is clearly 0.

Now consider $N(b)$. This is the number of solutions of (6.9) with $b \geq 3$. If we put $b = b' + 3$, (6.9) becomes $a + (b' + 3) + \ldots + z + \omega = 7$, i.e. $a + b' + \ldots + \omega = 4$, and so $N(b) = \binom{27+4-1}{4} = \binom{30}{4}$.

In the same way, we obtain

$$N(b) = N(c) = N(f) = N(h) = N(m) = N(p) = N(v) = N(w)$$
$$= N(y) = N(\omega) = \binom{30}{4},$$
$$N(j) = N(k) = N(q) = N(x) = N(z) = \binom{31}{5},$$
$$N(g) = \binom{29}{3}, \quad N(d) = N(l) = N(s) = N(u) = \binom{28}{2},$$
$$N(n) = N(r) = N(t) = \binom{26}{0} = 1, \quad \text{and all others are zero.}$$

We next have to deal with terms such as $N(c, d)$. There are $\binom{27}{2}$ such terms, but fortunately many are zero; for example $N(c, d)$ is zero since the requirements $c \geq 3$ and $d \geq 5$ are incompatible with (6.9).

There are $\binom{10}{2}$ terms equal to $N(b, c)$. Putting $b = b' + 3$ and $c = c' + 3$ leads to the equation $a + b' + c' + d + \ldots + \omega = 1$, which has $\binom{27}{1}$ solutions. Similarly there are 5 terms equal to $N(g, j)$. Putting $g = g' + 4$ and $j = j' + 2$ leads to the equation $a + \ldots + g' + h + i + j' + \ldots + = 1$, again with $\binom{27}{1}$ solutions. There are $\binom{5}{2} = 10$ terms equal to $N(j, k), 50 = 10 \times 5$ terms equal to $N(b, j)$; 10 like $N(b, g)$, 20 like $N(d, j)$.

Finally, there are $\binom{5}{3}$ terms equal to $N(j, k, q)$, and $\binom{5}{2} \times 10$ terms equal to $N(b, j, k)$. So the required number is

$$\binom{33}{7} - \left\{5\binom{31}{5} + 10\binom{30}{4} + \binom{29}{3} + 4\binom{28}{2} + 3\binom{26}{0}\right\}$$
$$+ \left\{\binom{5}{2}\binom{29}{3} + 50\binom{28}{2} + \binom{10}{2}\binom{27}{1} + 5\binom{27}{1} + 30\binom{26}{0}\right\}$$
$$- \left\{\binom{5}{3}\binom{27}{1} + 100\binom{26}{0}\right\}$$

$$= 3\,199\,724.$$

6.5 The Ménage Problem

The ménage problem was posed by the French mathematician E. Lucas in 1891.

Problem. In how many ways can n married couples sit round a table (with labelled seats) so that men and women alternate, and no husband and wife sit next to each other?

Solution

Let us be gentlemanly, and seat the ladies first. Observe that the wives $W_1, \ldots, W_n$ will occupy either the odd numbered seats or the even numbered seats; so the number of ways of seating the wives is twice $n!$. Then for each of these $2(n!)$ arrangements, there will be the same number $g(n)$ of ways of seating the husbands. The problem is now to evaluate $g(n)$. Without loss of generality, we can suppose that the wives are seated as shown in Figure 6.3,

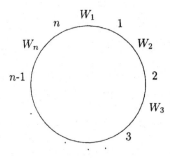

Figure 6.3

with the vacant seats now labelled $1, \ldots, n$, seat i being between wives W_i and W_{i+1} $(i < n)$, and seat n being between W_n and W_1.

Let S be the set of all possible ways of arranging the husbands in the n seats, so that $|S| = n!$. Then consider the following properties that elements of S may or may not possess:

$P_i : H_i$ is in seat i;

$Q_i : H_i$ is in seat $i - 1$ $(2 \leq i \leq n)$;

$Q_1 : H_1$ is in seat n.

Then $g(n)$ is just the number of elements of S possessing none of these properties. In applying (6.4), not every possible combination of properties is available; for example, P_1 and Q_2 cannot both be satisfied together. Properties which can be satisfied simultaneously are called **compatible**. Let r_k denote the number of ways of choosing k compatible properties from the P_i and the Q_i. Then, for each such choice, the number of arrangements of the husbands into seats so that these k properties are satisfied will be $(n - k)!$. Thus by (6.4),

$$g(n) = n! - r_1(n-1)! + r_2(n-2)! - \ldots + (-1)^n r_n 0!.$$

To find r_k, imagine the **properties** to be arranged in a circle as shown in Figure 6.4.

Properties are then compatible precisely when no two are adjacent; so r_k is just the number of ways of choosing k elements from a circular arrangement

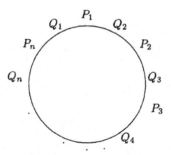

Figure 6.4

of $2n$ elements, no two of the chosen k being adjacent. Now it was shown in Exercise 2.15(b) that the number of ways of choosing k non-adjacent elements from $1, \ldots , 2n$ is $\binom{2n-k+1}{k}$; so r_k is this number minus the number of such choices which contain both 1 and $2n$. But if 1 and $2n$ are chosen then 2 and $2n-1$ are not, and so $k-2$ non-adjacent numbers are chosen from $3, \ldots , 2n-2$. Since there are $2n-4 = 2(n-2)$ numbers here, there are $\binom{2n-4-(k-2)+1}{k-2} = \binom{2n-k-1}{k-2}$ such choices. So finally

$$r_k = \binom{2n-k+1}{k} - \binom{2n-k-1}{k-2} = \frac{2n}{2n-k}\binom{2n-k}{k}$$

and the answer $M(n)$ to the ménage problem is $2(n!)g(n)$, where $g(n)$ is

$$n! - \frac{2n}{2n-1}\binom{2n-1}{1}(n-1)! + \frac{2n}{2n-2}\binom{2n-2}{2}(n-2)! - \ldots + (-1)^n \frac{2n}{n}\binom{n}{n}0!.$$

For example, $M(5) = 2 \times 5! \times 13 = 3120$.

Looking ahead, there is a nice alternative way of interpreting $g(n)$; it is the number of ways of choosing a third row of an $n \times n$ latin square whose first two rows are

$$
\begin{array}{cccccc}
1 & 2 & 3 & \cdots & n-1 & n \\
2 & 3 & 4 & \cdots & n & 1.
\end{array}
$$

Exercises

Exercise 6.1

Extend (6.2) to four sets A, B, C, D.

Exercise 6.2

Each student in a class of 100 reads at least one of mathematics and computing; 67 read mathematics and 44 read both. How many read computing?

Exercise 6.3

Each of the 100 students at a music school play at least one instrument, string, woodwind or brass. 70 play a string instrument, 49 woodwind and 49 brass. 20 play both string and woodwind; 25 play string and brass, and 35 play both woodwind and brass. How many play all three types of instrument?

Exercise 6.4

How many positive integers ≤ 1000 are divisible by none of $7, 11$ or 13?

Exercise 6.5

(a) Imitate the argument of Example 6.2 to show that if G is a graph with 100 vertices, each of degree > 75, then G must contain K_5 as a subgraph.

(b) Generalise to the following. If G is a graph with mn vertices, each of degree $> m(n-1)$, then G must contain K_{n+1} as a subgraph.

Exercise 6.6

Show that the number of solutions of the equation $x + y + z = 100$ in non-negative integers x, y, z with $x \leq 50, y \leq 40, z \leq 30$ is 231.

Exercise 6.7

How many permutations are there of $1, \ldots, 8$ in which none of the patterns $12, 34, 56, 78$ appears?

Exercise 6.8

How many permutations of $1, \ldots, 8$ are there in which no even number appears in its natural position?

Exercise 6.9

Repeat Example 6.5 on the Enigma machine, with 5 replaced by 8.

Exercise 6.10

How many permutations of $1, 1, 2, 2, 3, 3, 4, 4, 5, 5$ are there in which no two adjacent numbers are equal?

Exercise 6.11

The Euler phi function. Let $n = p_1^{\alpha_1} \dots p_r^{\alpha_r}$ be the prime factorisation of n and let $\phi(n)$ denote the number of positive integers $\leq n$ which are coprime to n (i.e. which are divisible by none of $p_1, \dots, p_r$). For example, $\phi(10) = 4$, the numbers being $1, 3, 7, 9$. Use the inclusion–exclusion principle to show that $\phi(n) = n \prod_i (1 - \frac{1}{p_i})$. Hence find $\phi(100), \phi(200)$.

Exercise 6.12

Let G be a graph with n vertices and m edges. Let S be the set of all possible colourings of the vertices using λ colours, ignoring adjacencies: so $|S| = \lambda^n$. For each $i \leq m$, let P_i denote the property that the endpoints of edge e_i receive the same colour. Then $f_\lambda(G)$, the number of vertex colourings of G using λ colours (see Exercise 5.19), is the number of elements of S possessing none of the properties P_i. Deduce that

$$f_\lambda(G) = \lambda^n + a_1 \lambda^{n-1} + a_2 \lambda^{n-2} + \cdots + a_{n-1} \lambda + a_n$$

where $a_1 = -m$ and $a_2 = \binom{m}{2} - t$, where t denotes the number of subgraphs of G isomorphic to K_3.

Exercise 6.13

(An example from Abraham de Moivre's book *Doctrines of Chance*, published in 1717.) If 12 throws of a die are made, what is the probability that all six numbers appear?

Exercise 6.14

Use the inclusion-exclusion principle to find the number of partitions of $\{1, \dots, 10\}$ into four parts, none of which is a singleton set.

7
Latin Squares and Hall's Theorem

In this chapter we study latin squares and their orthogonality, and construct magic squares. We also discuss Hall's theorem on systems of distinct representatives and apply it to latin squares and bipartite graphs. Finally, we show how complete sets of orthogonal latin squares lead to affine planes.

7.1 Latin Squares and Orthogonality

Definition 7.1

A latin square of order n is an $n \times n$ array in which each row and each column contains each of n given symbols exactly once.

Here, for example, are two latin squares of order 4, based on the set $\{1, 2, 3, 4\}$

$$
L_1 = \begin{bmatrix} 1 & 2 & 3 & 4 \\ 3 & 4 & 1 & 2 \\ 4 & 3 & 2 & 1 \\ 2 & 1 & 4 & 3 \end{bmatrix}, \quad L_2 = \begin{bmatrix} 1 & 2 & 3 & 4 \\ 4 & 3 & 2 & 1 \\ 2 & 1 & 4 & 3 \\ 3 & 4 & 1 & 2 \end{bmatrix}.
$$

Much interest in latin squares arose through their use in statistical experimental design, but they crop up in many different areas of discrete mathematics and algebra; a trivial example is the fact that the composition table of a finite group is a latin square.

Example 7.1

Suppose that $2n$ teams play in a league, with each team playing one game on each of $2n - 1$ consecutive weekends. Altogether each team has to play every other team once. Label the teams by $1, \ldots, 2n$, write $a_{ij} = k$ if teams i and j play each other on the kth weekend $(i \neq j)$ and put $a_{ii} = 2n$ for each i. Then $A = (a_{ij})$ is a latin square of order $2n$. For example, the games

$$\text{weekend } 1: \quad 1 \text{ v } 2, \quad 3 \text{ v } 4$$

$$\text{weekend } 2: \quad 1 \text{ v } 3, \quad 2 \text{ v } 4$$

$$\text{weekend } 3: \quad 1 \text{ v } 4, \quad 2 \text{ v } 3$$

give the latin square

$$\begin{bmatrix} 4 & 1 & 2 & 3 \\ 1 & 4 & 3 & 2 \\ 2 & 3 & 4 & 1 \\ 3 & 2 & 1 & 4 \end{bmatrix}.$$

Note that this latin square is **symmetric** about the main diagonal: $a_{ij} = a_{ji}$ for all i, j. Conversely, any symmetric latin square of order $2n$ with a constant main diagonal can be interpreted as the fixture list for a league of $2n$ teams.

Many applications of latin squares make use of the concept of **orthogonality**. This idea goes back to Euler. In 1779 he discussed a problem concerning 36 officers. These officers were from six different regiments, six from each; among the six from each regiment was one officer of each of six different ranks. Euler asked if it were possible to arrange the 36 officers in a 6×6 array, so that each row and each column contained one officer of each regiment and one of each rank. He believed (correctly, although he couldn't prove it) that such an arrangement is impossible.

Instead, consider the corresponding problem of 16 officers, from four regiments $\alpha, \beta, \gamma, \delta$, with one officer of each of the ranks a, b, c, d from each regiment. This problem **can** be solved. Here is a solution; in it, for example, γd stands for an officer from regiment γ and of rank d:

$$\begin{array}{cccc} \alpha a & \beta b & \gamma c & \delta d \\ \gamma d & \delta c & \alpha b & \beta a \\ \delta b & \gamma a & \beta d & \alpha c \\ \beta c & \alpha d & \delta a & \gamma b. \end{array}$$

Since there is to be one officer from each regiment in each row and column, the letters $\alpha, \beta, \gamma, \delta$ must form a latin square; similarly for a, b, c, d. Further, since there is just one officer of each rank from each regiment, the pairs (Greek letter, Latin letter) must all be distinct. This is achieved in the given solution, and the reader can check that the Greek letters correspond to the latin square L_1 and the Latin letters to the latin square L_2.

Definition 7.2

(i) If $A = (a_{ij})$ and $B = (b_{ij})$ are $n \times n$ arrays, the **join** (A, B) of A and B is the $n \times n$ array whose (i, j)th entry is the pair (a_{ij}, b_{ij}).

(ii) Latin squares A, B are **orthogonal** if all the entries in their join (A, B) are distinct.

If A and B are orthogonal, we call each of A and B an **orthogonal mate** of the other. Thus, for example, L_1 and L_2 are orthogonal mates. Note that the requirement that all the entries in (A, B) are distinct is equivalent to the requirement that each of the n^2 possible pairs occurs exactly once. Note also that the condition for orthogonality of A and B can be expressed as:

$$\text{if } a_{ij} = a_{IJ} \text{ and } b_{ij} = b_{IJ} \text{ then } i = I \text{ and } j = J. \tag{7.1}$$

The join of two latin squares is called a **Graeco-Latin square** because of Euler's use of Greek and Latin letters, as above. The name **latin square** for a single array was a later introduction. Euler's officer problem has a solution only if there exist two orthogonal latin squares of order 6; eventually, in 1900, it was proved beyond doubt that no such squares exist.

More generally, latin squares $A_1, \ldots, A_r$ of order n are **mutually orthogonal** if they are orthogonal in pairs, i.e. if, for all $i \neq j$, A_i and A_j are orthogonal. We shall use the abbreviation MOLS for mutually orthogonal latin squares.

For a given n, there is a limit to the number of MOLS of order n that can exist. We let $N(n)$ denote the largest value of r for which r MOLS of order n exist.

Theorem 7.1

$N(n) \leq n - 1$ for all $n \geq 2$.

Proof

Suppose $L_1, \ldots, L_r$ are r MOLS of order n. By relabelling the elements of each (which does not affect the orthogonality condition) we can suppose that each square has first row $1, 2, \ldots, n$. Concentrate on the entries in the $(2, 1)$ position. Since each square already has a 1 in the first column, none of these $(2, 1)$ entries can be 1. But, further, no two of them can be equal, for the join of any two of the squares already has each repeated pair in the first row. So $r \leq n - 1$.

A set of $n - 1$ MOLS of order n, if it exists, is called a **complete** set of MOLS.

Example 7.2

Here is a complete set of 3 MOLS of order 4:

$$
M_1 = \begin{bmatrix} 1 & 2 & 3 & 4 \\ 2 & 1 & 4 & 3 \\ 3 & 4 & 1 & 2 \\ 4 & 3 & 2 & 1 \end{bmatrix}, \quad M_2 = \begin{bmatrix} 1 & 2 & 3 & 4 \\ 4 & 3 & 2 & 1 \\ 2 & 1 & 4 & 3 \\ 3 & 4 & 1 & 2 \end{bmatrix}, \quad M_3 = \begin{bmatrix} 1 & 2 & 3 & 4 \\ 3 & 4 & 1 & 2 \\ 4 & 3 & 2 & 1 \\ 2 & 1 & 4 & 3 \end{bmatrix}.
$$

The next theorem establishes that complete sets exist wherever n is prime.

Theorem 7.2

$N(p) = p - 1$ for all primes p.

Proof

We define square arrays $A_1, \dots, A_{p-1}$ as follows. For the (i,j)th entry of A_k, take $a_{ij}^{(k)} = ki + j$, reduced modulo p to lie in the set $\{1, \dots, p\}$.

(i) We first check that each A_k is a latin square. First, the entries in the ith row are all different; for if $a_{ij}^{(k)} \equiv a_{iJ}^{(k)}$ then $ki + j \equiv ki + J (\bmod p)$ whence $j = J$. Secondly, the entries in the jth column are all distinct; for if $a_{ij}^{(k)} \equiv a_{Ij}^{(k)}$ then $ki + j \equiv kI + j (\bmod p)$ whence $k(i - I) \equiv 0 (\bmod p)$. Thus p divides $k(i - I)$, whence p divides $i - I$, so that $i \equiv I$ $(\bmod p)$ and i must equal I.

(ii) We now use (7.1) to check that A_k and A_h are orthogonal whenever $k \neq h$. Suppose $a_{ij}^{(k)} = a_{IJ}^{(k)}$ and $a_{ij}^{(h)} = a_{IJ}^{(h)}$. Then $ki+j \equiv kI+J$ and $hi+j \equiv hI+J (\bmod p)$. Subtracting one from the other gives $(h-k)i \equiv (h-k)I (\bmod p)$, and, as above, this gives $i = I$. Substituting back now gives $j \equiv J (\bmod p)$ whence $j = J$.

Example 7.3

Here is a complete set of 4 MOLS of order 5:

$$
A_1 = \begin{bmatrix} 2 & 3 & 4 & 5 & 1 \\ 3 & 4 & 5 & 1 & 2 \\ 4 & 5 & 1 & 2 & 3 \\ 5 & 1 & 2 & 3 & 4 \\ 1 & 2 & 3 & 4 & 5 \end{bmatrix}, \quad A_2 = \begin{bmatrix} 3 & 4 & 5 & 1 & 2 \\ 5 & 1 & 2 & 3 & 4 \\ 2 & 3 & 4 & 5 & 1 \\ 4 & 5 & 1 & 2 & 3 \\ 1 & 2 & 3 & 4 & 5 \end{bmatrix},
$$

$$A_3 = \begin{bmatrix} 4 & 5 & 1 & 2 & 3 \\ 2 & 3 & 4 & 5 & 1 \\ 5 & 1 & 2 & 3 & 4 \\ 3 & 4 & 5 & 1 & 2 \\ 1 & 2 & 3 & 4 & 5 \end{bmatrix}, A_4 = \begin{bmatrix} 5 & 1 & 2 & 3 & 4 \\ 4 & 5 & 1 & 2 & 3 \\ 3 & 4 & 5 & 1 & 2 \\ 2 & 3 & 4 & 5 & 1 \\ 1 & 2 & 3 & 4 & 5 \end{bmatrix}.$$

It should be noted here, for those readers who know about finite fields, that a similar type of argument establishes that $N(q) = q - 1$ whenever q is a prime power. (See Exercise 7.9.)

7.2 Magic Squares

A magic square of order n is an $n \times n$ array containing each of the numbers $1, \ldots, n^2$, and such that each row, each column and the two main diagonals all have a common sum (which in fact must be $\frac{1}{2}n(n^2 + 1)$; see Exercise 7.3).

Example 7.4

Here is a magic square of order 3:

$$\begin{bmatrix} 8 & 1 & 6 \\ 3 & 5 & 7 \\ 4 & 9 & 2 \end{bmatrix}.$$

Magic squares of all orders $n \geq 3$ exist. If n is odd, the method brought back from Siam by de la Loubère in the seventeenth century can be used, as in Example 7.4. Start with 1 in the centre of the top row; in general, travel northeast, and put the next number in the next square if the square is free; if you go off one edge, reappear at the opposite edge; if travelling northeast takes you to an occupied square, go south instead.

Example 7.5

Check that de la Loubère's method in the case $n = 5$ gives the following:

$$\begin{bmatrix} 17 & 24 & 1 & 8 & 15 \\ 23 & 5 & 7 & 14 & 16 \\ 4 & 6 & 13 & 20 & 22 \\ 10 & 12 & 19 & 21 & 3 \\ 11 & 18 & 25 & 2 & 9 \end{bmatrix}.$$

Methods of constructing magic squares of even order are more complicated. Some squares can be constructed by a general method due to Euler, involving MOLS.

Example 7.6

Take the join of the second and third latin squares of Example 7.2:

$$\begin{bmatrix} 11 & 22 & 33 & 44 \\ 43 & 34 & 21 & 12 \\ 24 & 13 & 42 & 31 \\ 32 & 41 & 14 & 23 \end{bmatrix}.$$

Reduce the first number in each pair by 1 to obtain

$$\begin{bmatrix} 01 & 12 & 23 & 34 \\ 33 & 24 & 11 & 02 \\ 14 & 03 & 32 & 21 \\ 22 & 31 & 04 & 13 \end{bmatrix} \qquad (7.2)$$

and then interpret the entries as "base 4" representations of the numbers 1 to 16 (xy standing for $4x + y$). This gives the following magic square:

$$\begin{bmatrix} 1 & 6 & 11 & 16 \\ 15 & 12 & 5 & 2 \\ 8 & 3 & 14 & 9 \\ 10 & 13 & 4 & 7 \end{bmatrix}.$$

Why this is a magic square is easy to see from (7.2); each row, column and main diagonal in (7.2) contains each of $0, 1, 2, 3$ once in the first position and each of $1, 2, 3, 4$ once in the second position, and therefore have equal sums.

This method will work in general provided we can find two MOLS in which both diagonals contain each element exactly once. It should now be clear why we did not take the first latin square of Example 7.2.

The following Indian magic square, dating back to the twelfth century, has the extra property that **every** (broken) diagonal has the same sum:

$$\begin{bmatrix} 7 & 12 & 1 & 14 \\ 2 & 13 & 8 & 11 \\ 16 & 3 & 10 & 5 \\ 9 & 6 & 15 & 4 \end{bmatrix}. \qquad (7.3)$$

For example, $12 + 8 + 5 + 9 = 12 + 2 + 5 + 15 = 34$. Such a magic square is called **pandiagonal** or diabolic. If we want to use Euler's method to construct diabolic magic squares, then we have to choose two orthogonal latin squares such that **every** broken diagonal has the same sum (or, in particular, contains all the elements exactly once). This can be achieved in many cases.

Example 7.7

Take A_2 and A_3 in Example 7.3. Reducing each element of A_2 by 1 and interpreting the pair (x, y) in the join by $5x + y$, we obtain

$$\begin{bmatrix} 14 & 20 & 21 & 7 & 13 \\ 22 & 3 & 9 & 20 & 21 \\ 10 & 11 & 17 & 23 & 4 \\ 18 & 24 & 5 & 6 & 12 \\ 1 & 7 & 13 & 19 & 25 \end{bmatrix}.$$

More generally, if n is odd and is not divisible by 3, we can take $A = (a_{ij})$ and $B = (b_{ij})$ where $a_{ij} \equiv 2i + j - 2 \pmod{n}, b_{ij} \equiv 3i + j - 2 \pmod{n}$. Then the join of A and B leads to a diabolic latin square. The details are left to the reader in Exercise 7.6.

7.3 Systems of Distinct Representatives

Latin squares can be constructed a row at a time. Given the first row, the second row has to be a derangement of the first; but, in general, if the first r rows of a hoped-for $n \times n$ latin square have been constructed, is it always possible to find a suitable $(r + 1)$th row? Since r elements have so far appeared in each column, the set X_i of available entries for the ith position in the $(r + 1)$th row has size $n - r$. The problem is: can a **different** element be chosen from each of $X_1, \ldots, X_n$? If so, then these elements will form an $(r + 1)$th row.

Example 7.8

Suppose we have the first 2 rows as follows:

$$\begin{bmatrix} 1 & 2 & 3 & 4 & 5 \\ 3 & 1 & 4 & 5 & 2 \\ . & & . & & \\ . & & . & & \\ . & & . & & \end{bmatrix}.$$

Here $X_1 = \{2,4,5\}, X_2 = \{3,4,5\}, X_3 = \{1,2,5\}, X_4 = \{1,2,3\}, X_5 = \{1,3,4\}$. We could choose $2,3,5,1,4$ from $X_1, \ldots, X_5$ respectively to get

$$
\begin{bmatrix}
1 & 2 & 3 & 4 & 5 \\
3 & 1 & 4 & 5 & 2 \\
2 & 3 & 5 & 1 & 4 \\
\cdot & & \cdot & & \\
\cdot & & & & \\
\cdot & & & &
\end{bmatrix}.
$$

Definition 7.3

A **system of distinct representatives** (SDR) for the sets $A_1, \ldots, A_m$ consists of **distinct** elements $x_1, \ldots, x_m$ such that $x_i \in A_i$ for each i.

Example 7.9

(a) $3,1,4,2$ form an SDR for the sets $\{1,3,5\}, \{1,2\}, \{3,4\}, \{2,3,4\}$.

(b) The sets $\{1,2,4\}, \{2,4\}, \{1,4\}, \{1,2\}, \{1,5\}, \{3,4,5\}$ do **not** possess an SDR since the first four of the sets contain only three elements in their union, not enough to provide **distinct** representatives of each.

This example illustrates the only situation which can stop a collection of sets having an SDR. We shall say that sets $A_1, \ldots, A_n$ satisfy the **Hall condition** if

$$
\begin{array}{c}
\text{for all } k \leq n, \text{ the union of any } k \text{ of the sets} \\
A_i \text{ contains at least } k \text{ elements.}
\end{array}
\tag{7.4}
$$

Theorem 7.3

The sets $A_1, \ldots, A_n$ possess an SDR if and only if they satisfy the Hall condition (7.4).

Proof

[We give a proof based on ideas of R. Rado. It uses the simplest form of the inclusion-exclusion principle: $|X \cup Y| = |X| + |Y| - |X \cap Y|$.]

If the sets possess an SDR they clearly must satisfy (7.4). So now suppose that $A_1, \ldots, A_n$ satisfy (7.4); we show they possess an SDR. We begin by removing, if possible, an element from one of the sets A_i so that the resulting sets still satisfy the Hall condition. We then continue to remove elements in this way, one at a time, until we obtain sets $B_1, \ldots, B_n$, with $B_i \subseteq A_i$ for each i, such that the removal of any further element from any B_i would cause the Hall condition for the B_i to be violated. If we can show that each B_i is a singleton set (i.e. $|B_i| = 1$ for each i), then the sets B_i themselves must be disjoint and hence provide an SDR for the A_i.

So suppose B_1 has two elements x, y. If either is removed, Hall's condition for the B_i is violated, so there are two sets P, Q of indices such that if

$$X = (B_1 - \{x\}) \cup \bigcup_{i \in P} B_i, \quad Y = (B_1 - \{y\}) \cup \bigcup_{i \in Q} B_i$$

then $|X| \leq |P|$ and $|Y| \leq |Q|$. But

$$X \cup Y = B_1 \cup \bigcup_{i \in P \cap Q} B_i, \quad X \cap Y \supseteq \bigcup_{i \in P \cap Q} B_i,$$

and the Hall condition gives

$$|X \cup Y| \geq 1 + |P \cup Q|, \quad |X \cap Y| \geq |P \cap Q|.$$

Thus the inclusion-exclusion principle gives

$$\begin{aligned} |P| + |Q| &\geq |X| + |Y| = |X \cup Y| + |X \cap Y| \\ &\geq 1 + |P \cup Q| + |P \cap Q| \\ &= 1 + |P| + |Q|, \end{aligned}$$

a contradiction. So B_1 must have just one element. A similar argument can be applied to each B_i, so $|B_i| = 1$ for all i, as required.

We now establish an important consequence of Theorem 7.3 (or **Hall's theorem**, as it is called), which will be particularly useful.

Theorem 7.4

Let $A_1, \ldots, A_n$ be subsets of S such that, for some m,
(i) $|A_i| = m$ for each i, and
(ii) each element of S occurs in exactly m of the A_i.
Then $A_1, \ldots, A_n$ possess an SDR.

Proof

We show that the sets A_i satisfy (7.4). Consider the union of k of the A_i. Including repetitions, the union contains km elements. But, by (ii), no element can occur in this union more than m times; so the number of distinct elements in the union is at least $\frac{km}{m} = k$.

The first use of Theorem 7.4 is to confirm that latin squares can be built up row by row.

Definition 7.4

If $r \leq n$, an $r \times n$ **latin rectangle** on an n-element set S is an $r \times n$ array of elements of S such that no element occurs more than once in any row or column.

Theorem 7.5

Any $r \times n$ latin rectangle with $r < n$ can be extended to an $(r + 1) \times n$ latin rectangle.

Proof

Let L be an $r \times n$ latin rectangle on $\{1, \ldots, n\}$. For each $i \leq n$ let A_i denote the set of elements of $\{1, \ldots, n\}$ which do not occur in the ith column of L. Then $|A_i| = n - r$ for each i. Further, for each $j \leq n$, j occurs in each row of L and hence has appeared in r of the columns; so j must occur in precisely $n - r$ of the A_i. Thus we can take $m = n - r$ in Theorem 7.4 and conclude that the sets A_i possess an SDR which can be taken as the $(r + 1)$th row of the required rectangle.

Our second application of Theorem 7.4 is really just a reformulation of the result in terms of bipartite graphs.

Definition 7.5

A set of disjoint edges in a graph G is called a **matching**. If G has $2n$ vertices a matching with n edges is called a **complete** or **perfect** matching of G.

Theorem 7.6

Let G be a bipartite graph with vertex set bipartition $V = B \cup W$ where $|B| = |W| = n$, and where every vertex of G has the same vertex degree m. Then G possesses a perfect matching.

Proof

Suppose that $B = \{u_1, \ldots, u_n\}$ and $W = \{v_1, \ldots, v_n\}$. For each $i \leq n$ let

$$A_i = \{j : u_i \text{ and } v_j \text{ are adjacent }\}.$$

Then the sets A_i satisfy the conditions of Theorem 7.4, and so the A_i possess an SDR. The SDR gives a perfect matching in which u_i is adjacent to v_j where j represents A_i.

For regular bipartite graphs such as those in Theorem 7.6 we immediately obtain an alternative proof of König's result that $\chi'(G) = \Delta$ for bipartite graphs. For we can first find a perfect matching and colour its edges with one colour. The graph obtained by removing this matching then satisfies the conditions of the theorem with $m = \Delta - 1$, so we can repeat the argument and obtain a second perfect matching which we then colour with another colour. Continuing in this way we partition the edge set into Δ perfect matchings and thus obtain an edge colouring using Δ colours.

Theorem 7.6 was in fact proved first by Steinitz in 1893 and then independently by König in 1914. It is, of course, a special case of Theorem 5.13, since the edges of any one colour form a matching.

7.4 From Latin Squares to Affine Planes

In this section we show how complete sets of MOLS give rise to an important family of designs.

We start with three MOLS of order four, as given in Example 7.2, along with the following "natural" array N:

$$N = \begin{bmatrix} 1 & 2 & 3 & 4 \\ 5 & 6 & 7 & 8 \\ 9 & 10 & 11 & 12 \\ 13 & 14 & 15 & 16 \end{bmatrix}.$$

From them, we construct 20 4-element sets as follows.

(a) The rows of N give four sets:

$$\{1,2,3,4\}, \; \{5,6,7,8\}, \; \{9,10,11,12\}, \; \{13,14,15,16\},$$

and the columns of N given another four sets:

$$\{1,5,9,13\}, \; \{2,6,10,14\}, \; \{3,7,11,15\}, \; \{4,8,12,16\}.$$

(b) The first of the three MOLS, namely M_1, gives four sets:

$$\{1,6,11,16\}, \; \{2,5,12,15\}, \; \{3,8,9,14\}, \; \{4,7,10,13\}.$$

Here the first set consists of the entries in N in the positions in which 1 appears in M_1, the second consists of the entries in n in the positions in which 2 appears in M_1; and so on.

(c) In the same way, M_2 gives rise to the following four sets:

$$\{1,8,10,15\}, \; \{2,7,9,16\}, \; \{3,6,12,13\}, \; \{4,5,11,14\}.$$

(d) Finally, M_3 similarly gives rise to:

$$\{1,7,12,14\}, \; \{2,8,11,13\}, \; \{3,5,10,16\}, \{4,6,9,15\}.$$

We now have 20 sets, each of size four; we call these sets **blocks**. The blocks possess an important property: no pair of elements from $\{1, \ldots, 16\}$ occurs in more than one of the blocks. Certainly, the blocks in (b), (c), (d) each contain one element from each row and each column of N, and so they will contain no pair in the same row or column of N; such pairs will occur precisely once in the blocks of (a). Consider now any pair not in the same row of column of N. Such a pair cannot occur in more than one block of (b), since the blocks in (b) are disjoint; and similarly it cannot occur in two blocks of (c) or of (d). Suppose a pair occurs in a block of (b) and a block of (c). Then there are two positions in which M_1 has the same entries and M_2 also has the same entries. But this contradicts orthogonality of M_1 and M_2. So we have indeed established that no pair occurs in more than one block.

Now each block contains $\binom{4}{2} = 6$ pairs, so the 20 blocks contain altogether $6 \times 20 = 120$ pairs. But the total number of pairs of elements of $\{1, \ldots, 16\}$ is $\binom{16}{2} = 120$; so **every** pair must occur in a block!

Thus the 20 blocks have the property that each pair of elements occurs in exactly one of the blocks. This is the **balance** property that is the basis of the study of balanced incomplete block designs which we shall look at in Chapter 9.

More generally, if we start with $n - 1$ MOLS $M_1, \ldots, M_{n-1}$ of order n on $\{1, \ldots, n\}$, and take N to be the $n \times n$ array with entries $1, 2, \ldots, n^2$ in order, we construct blocks as follows:

(α_1) n blocks from the rows of N

(α_2) n blocks from the columns of N

(β_1) n blocks from M_1, the ith block consisting of the entries in N in the positions in which i appears in M_1,

$\vdots$

(β_{n-1}) n blocks from M_{n-1}, similarly obtained.

This gives $(n + 1) \times n = n^2 + n$ blocks of size n, with the property that no two elements lie in more than one block. But these blocks contain altogether $n(n + 1)\binom{n}{2} = \frac{1}{2}n^2(n + 1)(n - 1) = \binom{n^2}{2}$ pairs; so **every** pair of elements of $\{1, \ldots, n^2\}$ occurs in precisely one block.

We thus obtain a collection of $n(n + 1)$ subsets (blocks) of a set of size n^2 such that

(i) each block contains n elements;

(ii) each element is in $n + 1$ blocks;

(iii) each pair of elements lies in exactly one block;

(iv) each pair of blocks intersect in at most one element.

(To check (ii), note simply that each element is in precisely one of the blocks of each of $\alpha_1, \alpha_2, \beta_1, \ldots, \beta_{n-1}$. To check (iv), suppose that blocks β_1, β_2 have two elements x, y in common. Then the pair $\{x, y\}$ would occur in two blocks, contradicting (iii)).

Such a system is called an **affine plane** of order n. Compare the elements with points, and the blocks with lines in ordinary geometry. The property (iii) corresponds to the fact that any two points determine a unique line, and (iv) corresponds to the fact that any two lines intersect in at most one point. Lines which do not intersect are usually called **parallel**; we can think of each of $(\alpha_1), \ldots, (\beta_{n-1})$ as consisting of a set of n parallel lines, each forming a partition of $\{1, \ldots, n^2\}$.

Example 7.10

Take the two MOLS of order 3 given by

$$M_1 = \begin{bmatrix} 1 & 2 & 3 \\ 2 & 3 & 1 \\ 3 & 1 & 2 \end{bmatrix}, \quad M_2 = \begin{bmatrix} 1 & 2 & 3 \\ 3 & 1 & 2 \\ 2 & 3 & 1 \end{bmatrix}$$

and take

$$N = \begin{bmatrix} 1 & 2 & 3 \\ 4 & 5 & 6 \\ 7 & 8 & 9 \end{bmatrix}.$$

Then we obtain 12 blocks of size 3:

$\{1, 2, 3\}, \{4, 5, 6\}, \{7, 8, 9\}, \quad \leftarrow$ rows of N

$\{1, 4, 7\}, \{2, 5, 8\}, \{3, 6, 9\}, \quad \leftarrow$ columns of N

$\{1, 6, 8\}, \{2, 4, 9\}, \{3, 5, 7\}, \quad \leftarrow$ from M_1

$\{1, 5, 9\}, \{2, 6, 7\}, \{3, 4, 8\}. \quad \leftarrow$ from M_2.

Each of these four sets of three blocks partitions $\{1, \ldots, 9\}$, and can be thought of as a set of "parallel lines". Any two elements occur together in a unique block. Thinking of the blocks as lines, we represent the set-up by Figure 7.1, where eight of the lines are drawn straight, and four are drawn as arcs. From orthogonal squares we have got to the verge of design theory. We shall take up the thread in the final chapter.

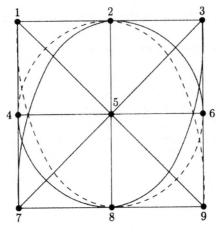

Figure 7.1

Exercises

Exercise 7.1

Write down two orthogonal squares of order 3 as given by Theorem 7.2.

Exercise 7.2

Write down the schedules of games given (as in Example 7.1) by the latin square

$$
\begin{bmatrix}
6 & 1 & 2 & 3 & 4 & 5 \\
1 & 6 & 5 & 4 & 2 & 3 \\
2 & 5 & 6 & 1 & 3 & 4 \\
3 & 4 & 1 & 6 & 5 & 2 \\
4 & 2 & 3 & 5 & 6 & 1 \\
5 & 3 & 4 & 2 & 1 & 6
\end{bmatrix}.
$$

Exercise 7.3

Verify that each row and each column of a magic square of order n must sum to $\frac{1}{2}n(n^2 + 1)$.

Exercise 7.4

Use de la Loubère's method to construct a magic square of order 7.

Exercise 7.5

Use Euler's method to construct a diabolic magic square of order 7.

Exercise 7.6

Verify that if n is odd and not divisible by 3, then $A = (a_{ij})$ and $B = (b_{ij})$, where $a_{ij} \equiv 2i + j - 2, b_{ij} \equiv 3i + j - 2 \pmod{n}$, are orthogonal latin squares of order n and that, in both, every diagonal contains each of $1, \ldots, n$.

Exercise 7.7

A latin square A is **self-orthogonal** if it is orthogonal to its transpose A^T. (This is the usual matrix transpose.)
(a) Verify that, in Example 7.2, M_2 is self-orthogonal.
(b) Are M_1, M_3 self-orthogonal?
(c) Show that, if $(n, 6) = 1$, the latin square A of Exercise 7.6 is self-orthogonal.

Exercise 7.8

(a) A symmetric latin square on $\{1, \ldots, n\}$ has $1, \ldots, n$ in its main diagonal. Show that n must be odd.
(b) Let $n = 2m + 1$ and define a_{ij} to be the member of $\{1, \ldots, n\}$ for which $a_{ij} \equiv (i + j)(m + 1) \pmod{n}$. Show that $A = (a_{ij})$ is a symmetric latin square with $1, \ldots, n$ on the main diagonal. Find A in the case $n = 5$.

Exercise 7.9

Show that $N(q) = q - 1$ for all prime power q, using the finite field GF(q). Let GF(q) = $\{\lambda_1, \lambda_2, \ldots, \lambda_{q-1}, \lambda_q = 0\}$, and define the squares A_k, $1 \leq k \leq q - 1$, by $a_{ij}^{(k)} = \lambda_i \lambda_k + \lambda_j$. (Imitate the proof of Theorem 7.2.)

Exercise 7.10

Apply Euler's diabolic square construction method to the square M_2 of Example 7.2 and its transpose M_2^T. Compare your answer with (7.3).

Exercise 7.11

Extend $\begin{bmatrix} 1 & 2 & 3 & 4 & 5 \\ 4 & 3 & 1 & 5 & 2 \end{bmatrix}$ to a latin square of order 5.

Exercise 7.12

Find an SDR for the sets $\{1,3,5\}, \{1,4,5\}, \{2,3,4\}, \{1,2,4\}$.

Exercise 7.13

Explain why the sets $\{1,2,3,4\}, \{2,5,6\}, \{1,4,5\}, \{2,6\}, \{5,8\}, \{1,4,7\}, \{2,5\}, \{5,6\}$ do not possess an SDR.

Exercise 7.14

The 52 cards of an ordinary pack, consisting of four suits of 13 different values, are arranged in a 4×13 array. Prove that 13 cards of different values, one from each column, can be chosen.

Exercise 7.15

A set S of mn elements is partitioned into m sets of size n in two different ways: $S = A_1 \cup \ldots \cup A_m = B_1 \cup \ldots \cup B_m$. Show that the sets B_i can be relabelled so that $A_i \cap B_i \neq \emptyset$ for each $i = 1, \ldots, m$. [Hint: consider the sets $S_i = \{j : A_i \cap B_j \neq \emptyset\}$.]

Exercise 7.16

Starting with the four MOLS of order 5 given in Example 7.3, construct an affine plane of order 5.

Exercise 7.17

An $n \times n$ matrix is called a **permutation matrix** if all entries are 0 or 1, and there is precisely one 1 in each row and each column. Show that, if M is an $n \times n$ matrix with all entries 0 or 1, with exactly m 1s in each row and column, then M can be written as the sum of m permutation matrices. Illustrate in the case where M is the matrix given by (9.2) in Chapter 9.

8

Schedules and 1-Factorisations

This chapter deals with the construction of league schedules for sports competitions or, equivalently, experimental designs involving comparisons of pairs of varieties. The subject matter provides nice connections with latin squares and edge colourings of graphs, and also provides an introduction to the ideas of block designs and resolvability which will be studied further in the final chapter. So the chapter is an exposition of interrelations between apparently different combinatorial ideas.

8.1 The Circle Method

Suppose that a football league contains eight teams, each of which plays every other team once. The games are to be arranged for seven Saturdays, with four games each Saturday, each team playing in one game each Saturday. How can a fixture list be constructed?

Suppose also that a biology researcher wishes to compare eight types of treatment, comparing each pair of treatments once. During the first week she will carry out four comparisons, using all eight treatments; then during the second week another four comparisons, and so on. Construct a suitable schedule of comparisons for seven weeks.

These two problems are clearly equivalent. Indeed, they are both equivalent to partitioning the set of edges of K_8 into seven sets of four disjoint edges (i.e. into seven complete matchings), four edges for each of seven weeks. If we replace weeks by colours, we see that both problems are equivalent to finding an edge colouring of K_8 using seven colours.

Example 8.1

The edge colouring of K_4 shown in Figure 5.4(a) shows a 3-colouring using colours $1, 2, 3$. It corresponds to the league schedule:

$$\begin{array}{lll} \text{round 1}: & A \text{ v } B, & C \text{ v } D \\ \text{round 2}: & A \text{ v } C, & B \text{ v } D \\ \text{round 3}: & A \text{ v } D, & B \text{ v } C. \end{array}$$

For $2n$ teams, we have to find an edge colouring of K_{2n} using $2n - 1$ colours. We have already seen how to do this in Theorem 5.11. We describe the method again, in slightly different language.

The circle method

Label the teams by $\infty, 1, 2, \ldots, 2n - 1$, and place $1, 2, \ldots, 2n - 1$ equally spaced round a circle with ∞ at the centre. The case $2n = 8$ is shown in Figure 8.1. In Figure 8.1(a) we have the fixtures for the first day, and Figure 8.1(b) gives the fixtures for the second day. By rotating the chords we obtain the games

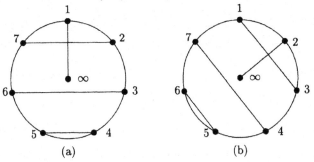

$$\begin{array}{cc} \text{(a)} & \text{(b)} \end{array}$$

Figure 8.1

for each of the $2n - 1$ days. In general, for $2n$ teams $\infty, 1, \ldots, 2n - 1$, on day i we play the games

$$i \text{ v } \infty, (i - 1) \text{ v } (i + 1), (i - 2) \text{ v } (i + 2), \ldots, (i - (n - 1)) \text{ v } (i + (n - 1))$$

where each integer is reduced (mod $2n - 1$) to lie in the set $\{1, \ldots, 2n - 1\}$.

Example 8.2

Fixture list for eight teams constructed by the circle method.

$$\begin{array}{lllll} \text{Day 1} & \infty \text{ v } 1 & 7 \text{ v } 2 & 6 \text{ v } 3 & 5 \text{ v } 4 \\ \text{Day 2} & \infty \text{ v } 2 & 1 \text{ v } 3 & 7 \text{ v } 4 & 6 \text{ v } 5 \\ \text{Day 3} & \infty \text{ v } 3 & 2 \text{ v } 4 & 1 \text{ v } 5 & 7 \text{ v } 6 \\ \text{Day 4} & \infty \text{ v } 4 & 3 \text{ v } 5 & 2 \text{ v } 6 & 1 \text{ v } 7 \\ \text{Day 5} & \infty \text{ v } 5 & 4 \text{ v } 6 & 3 \text{ v } 7 & 2 \text{ v } 1 \\ \text{Day 6} & \infty \text{ v } 6 & 5 \text{ v } 7 & 4 \text{ v } 1 & 3 \text{ v } 2 \\ \text{Day 7} & \infty \text{ v } 7 & 6 \text{ v } 1 & 5 \text{ v } 2 & 4 \text{ v } 3 \end{array}$$

Note that, on day i, teams u and v play each other where $u+v \equiv 2i \pmod{2n-1}$. Note too that the schedule is **cyclic** in the sense that each round is obtained by adding $1 \pmod{2n-1}$ to each entry in the previous round (∞ earning itself its name: $\infty + 1 = \infty$).

The circle method yields a partition of the edge-set of K_{2n} into n edge disjoint perfect matchings (1-factors). The name 1-factor reminds us that each vertex in a perfect matching has degree 1 in that matching. For example, the edge colouring of K_4 in Figure 5.4(a) gives a partition of the edge-set of K_4 into three 1-factors.

Definition 8.1

A 1-factorisation of a graph G with $2n$ vertices is a partition of the edge-set into 1-factors.

Thus the circle method establishes the following result.

Theorem 8.1

K_{2n} has a 1-factorisation.

Not all graphs with an even number of vertices have 1-factorisations. Indeed, it should be clear that a graph can have a 1-factorisation only if it is **regular**, i.e. if every vertex has the same degree. If each vertex has degree r, then the 1-factorisation, if it exists, will consist of r 1-factors. Since a 1-factorisation clearly gives an edge colouring of the graph with the minimum possible number of colours, and, conversely, an edge-colouring of a graph, regular of vertex degree Δ, using Δ colours, clearly gives a 1-factorisation, we have:

Theorem 8.2

A regular graph G with $2n$ vertices has a 1-factorisation if and only if it is class 1.

Thus, by Example 5.13, the Petersen graph does not have a 1-factorisation.

Example 8.3

How many 1-factorisations does K_6 have? Let K_6 have vertices $a, \dots, f$, and suppose that one of the 1-factors in a 1-factorisation is ab, cd, ef. In some other 1-factor we must have edge ac. There are only two possibilities:

$$ac, be, df \quad \text{or} \quad ac, bf, de.$$

But the 1-factorisation including

$$\begin{array}{lll} ab & cd & ef \\ ac & be & df \\ ad \\ ae \\ af \end{array}$$

can be completed in only one way (check this!) as can

$$\begin{array}{lll} ab & cd & ef \\ ac & bf & de \\ ad \\ ae \\ af \end{array}$$

So there are only two 1-factorisations containing the 1-factor ab, cd, ef. Similarly there are only two 1-factorisations containing any other 1-factor. But the total number of 1-factors is just the number of ways of partitioning a set of six elements into three pairs, and so, by Corollary 5.2, is $\frac{6!}{2^3.3!} = 15$. For each of the 15 choices of 1-factor, there are two ways of extending it to a 1-factorisation, giving 30 1-factorisations altogether. But each distinct 1-factorisation arises five times in this way, depending on which of the five 1-factors in it is taken first. So the number of **distinct** 1-factorisations is $\frac{30}{5} = 6$.

Thus K_6 has six different 1-factorisations. The number of 1-factorisations of K_{2n} rises dramatically with n: for K_8 there are 6240, and for K_{10} there are 1 255 566 720.

Schedules for $2n + 1$ teams

If a league schedule is to be arranged for $2n+1$ teams, then there can be at most n games on any one day, with one team resting. Such a schedule corresponds to an edge colouring of K_{2n+1} using $2n + 1$ colours, as in Theorem 5.11. The easiest way to obtain such a schedule is use the circle method to construct a schedule for $2n + 2$ teams and then omit all the games involving ∞.

Example 8.4

Suppose we want a schedule for five teams. Take the schedule for six teams given by the circle method:

$$\begin{array}{lll} \infty \text{ v } 1 & 5 \text{ v } 2 & 3 \text{ v } 4 \\ \infty \text{ v } 2 & 1 \text{ v } 3 & 4 \text{ v } 5 \\ \infty \text{ v } 3 & 2 \text{ v } 4 & 5 \text{ v } 1 \\ \infty \text{ v } 4 & 3 \text{ v } 5 & 1 \text{ v } 2 \\ \infty \text{ v } 5 & 4 \text{ v } 1 & 2 \text{ v } 3. \end{array}$$

Omit all the games involving ∞, to obtain

$$
\begin{array}{ll}
5\,\text{v}\,2 & 3\,\text{v}\,4 \\
1\,\text{v}\,3 & 4\,\text{v}\,5 \\
2\,\text{v}\,4 & 5\,\text{v}\,1 \\
3\,\text{v}\,5 & 1\,\text{v}\,2 \\
4\,\text{v}\,1 & 2\,\text{v}\,3.
\end{array}
$$

This is in fact the schedule used in recent years for the Five Nations Rugby championship involving England, France, Ireland, Scotland and Wales. For example, in the final year of that championship (1999), the key was:

$1 =$ England, $2 =$ France, $3 =$ Scotland, $4 =$ Wales, $5 =$ Ireland.

In the year 2000 the championship became the Six Nations championship with the inclusion of Italy. The schedule used was that obtained above for six teams, with the first round ∞ v $1, 2$ v $5, 4$ v 3 and with the games involving ∞ being given suitable home and away orientations. This gave

$$
\begin{array}{lll}
\infty\,\text{v}\,1 & 2\,\text{v}\,5 & 4\,\text{v}\,3 \\
2\,\text{v}\,\infty & 3\,\text{v}\,1 & 5\,\text{v}\,4 \\
3\,\text{v}\,\infty & 4\,\text{v}\,2 & 1\,\text{v}\,5 \\
\infty\,\text{v}\,4 & 5\,\text{v}\,3 & 2\,\text{v}\,1 \\
5\,\text{v}\,\infty & 1\,\text{v}\,4 & 3\,\text{v}\,2
\end{array}
$$

with the key $\infty =$ Italy, $1 =$ Scotland, $2 =$ Wales, $3 =$ Ireland, $4 =$ England, $5 =$ France. In the Five Nations championship the schedule automatically arranged for each team to alternate home and away games. This is not possible for an even number of teams (see Exercise 8.10); in the above, each team has one "break" in the alternating of venues.

Summing up, we have the following theorem.

Theorem 8.3

(i) For all $n \geq 1$, there exists an arrangement of the $\binom{2n}{2} = n(2n-1)$ 2-element subsets of $\{1, \dots, 2n\}$ into $2n - 1$ classes, each class consisting of n disjoint pairs.

(ii) For all $n \geq 1$, there exists an arrangement of the $\binom{2n+1}{2} = n(2n + 1)$ 2-element subsets of $\{1, \dots, 2n + 1\}$ into $2n + 1$ classes, each consisting of n disjoint pairs, with each element being absent from the pairs in exactly one class.

There is a natural generalisation of (i). Can the $\binom{3n}{3} = \frac{1}{2}n(3n - 1)(3n - 2)$ 3-element subsets (triples) of $\{1, \dots, 3n\}$ be arranged into $\frac{1}{2}(3n - 1)(3n - 2)$ classes each consisting of n disjoint triples? The case $n = 2$ is trivial, since every triple can be paired with its complement. The case $n = 3$ was solved by Sylvester. Here is an elegant solution.

Example 8.5

To put the 84 3-element subsets of $\{1, \ldots, 9\}$ into 28 groups of three triples such that the triples in each group form a partition of $\{1, \ldots, 9\}$.

Solution
Consider the following seven square arrays.

$$
\begin{array}{ccccccc}
123 & 372 & 234 & 318 & 531 & 912 & 261 \\
456 & 156 & 756 & 426 & 486 & 453 & 459 \\
789 & 489 & 189 & 759 & 729 & 786 & 783
\end{array}
$$

For each array we get four groups of 3 triples, from the rows, the columns, the forward leading "diagonals" and the backward leading diagonals. The first array gives

123,	456,	789	(rows)
147,	258,	369	(columns)
159,	267,	348	(forward diagonals)
168,	249,	357	(backward diagonals).

Similarly we obtain four partitions of $\{1, \ldots, 9\}$ from each of the other six arrays. Altogether we obtain 28 groups of three triples, containing between them all the 84 3-element subsets.

There is a remarkable theorem due to Baranyai (1973) which we state without proof. A proof can be found in [18].

Theorem 8.4

The set of all $\binom{nk}{k}$ k-element subsets of $\{1, \ldots, nk\}$ can be partitioned into $\frac{1}{n}\binom{nk}{k} = \binom{nk-1}{k-1}$ classes, each class consisting of n disjoint k-element subsets.

8.2 Bipartite Tournaments and 1-Factorisations of $K_{n,n}$

Example 8.6

Two schools, Alpha Academy and Beta High School, arrange a tennis match in which each school is represented by 4 players. Each player is to play each player in the opposing team once, and the games are to be arranged in four rounds, everyone playing in each round. Arrange a schedule.

Solution

Let the players of Alpha Academy be $A_1, \ldots, A_4$, and those of Beta High School be $B_1, \ldots, B_4$.

Round 1 $A_1 \text{ v } B_1, A_2 \text{ v } B_2, A_3 \text{ v } B_3, A_4 \text{ v } B_4$

Round 2 $A_1 \text{ v } B_2, A_2 \text{ v } B_3, A_3 \text{ v } B_4, A_4 \text{ v } B_1$

Round 3 $A_1 \text{ v } B_3, A_2 \text{ v } B_4, A_3 \text{ v } B_1, A_4 \text{ v } B_2$

Round 4 $A_1 \text{ v } B_4, A_2 \text{ v } B_1, A_3 \text{ v } B_2, A_4 \text{ v } B_3$.

Note that this schedule can be represented by a latin square M: in the ith column of M list the subscripts of the opponents of A_i in order:

$$M = \begin{bmatrix} 1 & 2 & 3 & 4 \\ 2 & 3 & 4 & 1 \\ 3 & 4 & 1 & 2 \\ 4 & 1 & 2 & 3 \end{bmatrix}.$$

Conversely, any latin square on $1, \ldots, 4$ can be interpreted as a schedule for a bipartite tournament by reversing this process. Thus bipartite tournaments (i.e. tournaments between two teams in which each player of one team plays every player of the other team) are equivalent to latin squares.

Note also that the solution to Example 8.6 can be re-expressed in terms of a 1-factorisation of the graph $K_{4,4}$. In Figure 8.2, the edges of $K_{4,4}$ are coloured using four colours, corresponding to the rounds in which the games (represented by edges) are played. Thus a 1-factorisation of $K_{n,n}$ is equivalent to a bipartite tournament for two teams of size n.

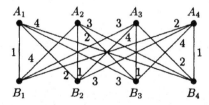

Figure 8.2

Since there is a latin square of order n for all $n \geq 1$, we have the following result.

Theorem 8.5

For all $n \geq 1, K_{n,n}$ has a 1-factorisation.

There is another way of representing a bipartite tournament by a latin square. Define the latin square $N = (n_{ij})$ by

$$n_{ij} = k \text{ if } A_i \text{ plays } B_j \text{ in round } k.$$

For the schedule in Example 8.5 we obtain

$$N = \begin{bmatrix} 1 & 2 & 3 & 4 \\ 4 & 1 & 2 & 3 \\ 3 & 4 & 1 & 2 \\ 2 & 3 & 4 & 1 \end{bmatrix}.$$

The square N is **conjugate** to the square $M : n_{ij} = k \Leftrightarrow m_{ki} = j$. Other conjugates of M are obtained similarly by permuting i, j, k. Note that the transpose M^T of M is a conjugate of $M : m_{ij}^T = k \Leftrightarrow m_{ji} = k$.

Another 1-factorisation of K_{2n}

The existence of a 1-factorisation of $K_{n,n}$ enables us to give another general method of constructing a 1-factorisation of K_{2n}, different from the one arising from the circle method. We describe it in terms of a league schedule.

Given $2n$ teams, label them $x_1, \ldots, x_n, y_1, \ldots, y_n$. We can then construct a bipartite tournament involving n rounds, in the course of which every x_i plays every y_i once. Without loss of generality, we can assume that, for each i, x_i plays y_i in the first round.

We now consider two cases.

(i) If n is **even**, we can then play a league of $n-1$ rounds involving the x_i in parallel with a league involving the y_i. This will give $n-1$ further rounds, thus completing the required tournament. Here is an example with $n = 4$.

Example 8.7

$x_1 \vee y_1$	$x_2 \vee y_2$	$x_3 \vee y_3$	$x_4 \vee y_4$
$x_1 \vee y_2$	$x_2 \vee y_3$	$x_3 \vee y_4$	$x_4 \vee y_1$
$x_1 \vee y_3$	$x_2 \vee y_4$	$x_3 \vee y_1$	$x_4 \vee y_2$
$x_1 \vee y_4$	$x_2 \vee y_1$	$x_3 \vee y_2$	$x_4 \vee y_3$
$x_1 \vee x_2$	$x_3 \vee x_4$	$y_1 \vee y_2$	$y_3 \vee y_4$
$x_1 \vee x_3$	$x_2 \vee x_4$	$y_1 \vee y_3$	$y_2 \vee y_4$
$x_1 \vee x_4$	$x_2 \vee x_3$	$y_1 \vee y_4$	$y_2 \vee y_3$

(ii) If n is **odd**, this idea does not work; the bipartite tournament cannot be extended to a full schedule since the games involving the odd number of x_i require n further rounds. But we can adapt the idea. Introduce two new teams ∞_x, ∞_y, and use the circle construction to obtain two schedules S_x, S_y on $\{\infty_x, x_1, \dots, x_n\}$ and $\{\infty_y, y_1, \dots, y_n\}$ respectively, in which ∞_x plays x_i and ∞_y plays y_i in round i. For the ith round of the required schedule, take the ith round games of S_x and S_y, but replace the games $\infty_x \vee x_i$ and $\infty_y \vee y_i$ by the one game $x_i \vee y_i$. The resulting n rounds, along with all but the first round of the bipartite tournament, give the required schedule.

Example 8.8

We construct a schedule for $n = 3$. Here S_x is

$$\begin{array}{ll} \infty_x \vee x_1, & x_2 \vee x_3 \\ \infty_x \vee x_2, & x_1 \vee x_3 \\ \infty_x \vee x_3, & x_1 \vee x_2 \end{array}$$

and a bipartite tournament is

$$\begin{array}{lll} x_1 \vee y_1, & x_2 \vee y_2, & x_3 \vee y_3 \\ x_1 \vee y_2, & x_2 \vee y_3, & x_3 \vee y_1 \\ x_1 \vee y_3, & x_2 \vee y_1, & x_3 \vee y_2. \end{array}$$

The final schedule is

$$\begin{array}{lll} x_1 \vee y_1, & x_2 \vee x_3, & y_2 \vee y_3 \\ x_2 \vee y_2, & x_1 \vee x_3, & y_1 \vee y_3 \\ x_3 \vee y_3, & x_1 \vee x_2, & y_1 \vee y_2 \\ x_1 \vee y_2, & x_2 \vee y_3, & x_3 \vee y_1 \\ x_1 \vee y_3, & x_2 \vee y_1, & x_3 \vee y_2. \end{array}$$

8.3 Tournaments from Orthogonal Latin Squares

Court balance

Suppose now that, in the tennis match of Example 8.6, four courts are available, of differing quality, and it is requested that the games should be arranged so that, not only does every A_i play every B_j once, but each player plays once on each court.

One way to solve this problem is to take the join of two MOLS of order 4, say

$$\begin{bmatrix} 11 & 22 & 33 & 44 \\ 24 & 13 & 42 & 31 \\ 32 & 41 & 14 & 23 \\ 43 & 34 & 21 & 12 \end{bmatrix}.$$

Take the rows as corresponding to the rounds, and the columns as corresponding to the courts. The latin property ensures that each player plays once in each round and once on each court; orthogonality ensures that every A_i plays every B_j once. So we obtain the solution given in Table 8.1.

Table 8.1

	Court 1	Court 2	Court 3	Court 4
Round 1	A_1 v B_1	A_2 v B_2	A_3 v B_3	A_4 v B_4
Round 2	A_2 v B_4	A_1 v B_3	A_4 v B_2	A_3 v B_1
Round 3	A_3 v B_2	A_4 v B_1	A_1 v B_4	A_2 v B_3
Round 4	A_4 v B_3	A_3 v B_4	A_2 v B_1	A_1 v B_2

Mixed doubles

Another use of orthogonal latin squares is in the construction of mixed doubles tournaments. Suppose that Alpha Academy and Beta High School now decide to play mixed doubles: each school provides four boys and four girls, and each player is to play in four games, partnering each person of the opposite sex once, and opposing each player of the other team once. The 16 games have to be arranged in four rounds, each of four games, each player being involved in one game in each round.

Let us denote Alpha's boys by $B_1, \ldots, B_4$ and Beta's boys by $b_1, \ldots, b_4$; similarly we denote Alpha's girls by $G_1, \ldots, G_4$ and Beta's by $g_1 \ldots, g_4$. Take also the MOLS M_1, M_2, M_3 of Example 7.2.

We interpret M_1 as giving the partners of the Alpha boys: if the (i,j)th entry is k, then B_i partners G_k when playing against b_j. Similarly M_2 gives the partners of the Beta boys: if the (i,j)th entry of M_2 is ℓ then b_j partners g_ℓ when playing against B_i. Since M_1, M_2 are latin squares, no repetition of partners occurs. Since the squares are orthogonal, no girl opposes another girl more than once.

Having obtained the games of the required schedule, it now remains to arrange them into four rounds. This is achieved by using M_3: if the (i,j)th entry of M_3 is k, place the game involving B_i and b_j in round k. Suppose this resulted in girl G_i playing two games in round k. Then we would have two games $B_r G_i$ v $b_s g_t$ and $B_u G_i$ v $b_x g_y$ in round k. But then the (r,s)th and (u,x)th entries of M_3 are both k and the (r,s)th and (u,x)th entries of M_1 are both i, contradicting orthogonality of M_1 and M_3. Similarly the orthogonality of M_2 and M_3 prevents any girl g_i playing twice in one round.

Thus, for example, since the $(3,2)$ entries of M_1 and M_2 are 4 and 1, one of the games will be $B_3G_4 \vee b_2g_1$. Since M_3 has 3 in the $(3,2)$ position, we assign this game to round 2. In this way we obtain the following schedule:

Round 1	$B_1G_1 \vee b_1g_1$	$B_2G_4 \vee b_3g_2$	$B_3G_2 \vee b_4g_3$	$B_4G_3 \vee b_2g_4$
Round 2	$B_1G_2 \vee b_2g_2$	$B_2G_3 \vee b_4g_1$	$B_3G_1 \vee b_3g_4$	$B_4G_4 \vee b_1g_3$
Round 3	$B_1G_3 \vee b_3g_3$	$B_2G_2 \vee b_1g_4$	$B_3G_4 \vee b_2g_1$	$B_4G_1 \vee b_4g_2$
Round 4	$B_1G_4 \vee b_4g_4$	$B_2G_1 \vee b_2g_3$	$B_3G_3 \vee b_1g_2$	$B_4G_2 \vee b_3g_1$

Exercises

Exercise 8.1

Use the circle method to construct a league schedule for 10 teams. Deduce a league schedule for 9 teams.

Exercise 8.2

Use the method of Example 8.8 to construct a league schedule for 10 teams.

Exercise 8.3

Suppose that the first r rounds of a bipartite tournament between two teams of size n have been constructed $(r < n)$. Can these r rounds always be extended to a full bipartite tournament?

Exercise 8.4

Find a 1-factor, if one exists, for (a) the Petersen graph, (b) the ethane graph (Figure 3.4), (c) the graph of an octahedron, (d) the graph of a cube.

Exercise 8.5

Partition the edges of the graph of an octahedron into two disjoint hamiltonian cycles. Deduce that the graph has a 1-factorisation.

Exercise 8.6

Which of the platonic graphs have a 1-factorisation?

Exercise 8.7

Prove that any hamiltonian graph in which each vertex has degree 3 has a 1-factorisation.

Exercise 8.8

(a) Imitate Example 8.6 for teams of 5 players.
(b) Obtain a schedule in which each player plays once on each of 5 courts.

Exercise 8.9

(a) Construct a mixed doubles tournament between Gamma Grammar School and Delta District Comprehensive, with 5 boys and 5 girls in each team.
(b) Suppose there are 5 courts available. Construct a schedule in which each player plays exactly once on each court.

Exercise 8.10

In a league schedule for $2n$ teams it is desired that each team should alternate home and away fixtures as much as possible. A repetition of home (or away) fixtures in two consecutive games is called a **break**. For example, in Example 8.2, team 6 plays $HHAAAAH$ and so has 4 breaks.

(a) Show that any schedule for $2n$ teams can have at most two teams with no breaks, and deduce that there must be at least $2n-2$ breaks altogether.
(b) Show that a schedule with exactly $2n-2$ breaks can be constructed. (In Figure 8.1, take the home teams as those alternately at left, right ends of chords, and alternate ∞ home and away throughout.)

Exercise 8.11

Suppose a league schedule has been constructed for $2n$ teams, but venues have not yet been assigned. Show that it is possible to assign venues so that, in each round, precisely one of the two teams of each first round game is at home.

9
Introduction to Designs

We introduce the idea of a balanced incomplete block design, and look at some special families of such designs, namely finite projective planes, affine planes, Steiner triple systems and Hadamard designs. The connection between finite projective planes and complete sets of MOLS is established. We also describe the usefulness of difference systems in the construction of designs. Finally we give a brief introduction to some of the ideas behind error-correcting codes.

9.1 Balanced Incomplete Block Designs

Example 9.1

Seven golfers are to spend a week's holiday at a seaside town which boasts two splendid golf courses. They decide that each should play a round of golf on each of the seven days. They also decide that on each day they should split into two groups, one of size 3 to play on one course, and the other of size 4 to play on the other course. Can the groups be arranged so that each pair of golfers plays together in a group of 3 the same number of times, and each pair plays together in a group of 4 the same number of times?

Solution

Here is one solution: the groups for each day are shown. It can be easily checked that each pair plays together once in a group of 3 and twice in a group of 4.

Day 1 $\{1,2,4\}$ $\{3,5,6,7\}$
Day 2 $\{2,3,5\}$ $\{4,6,7,1\}$
Day 3 $\{3,4,6\}$ $\{5,7,1,2\}$
Day 4 $\{4,5,7\}$ $\{6,1,2,3\}$
Day 5 $\{5,6,1\}$ $\{7,2,3,4\}$
Day 6 $\{6,7,2\}$ $\{1,3,4,5\}$
Day 7 $\{7,1,3\}$ $\{2,4,5,6\}$

What we have done is to make use of the configuration known as the **seven-point plane**, shown in Figure 9.1. In it, there are seven points and seven lines, with each line containing three points, and each pair of points being in exactly one line. The groups of size 4 are the **complements** of the lines of this configuration.

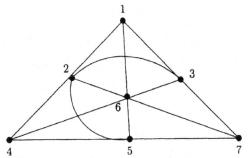

Figure 9.1 The seven-point plane

Example 9.2

The following subsets of $\{1,\dots,6\}$ have the property that each subset has 3 elements and each pair of elements occurs in two of the subsets:

$$\{1,2,3\},\{1,2,4\},\{1,3,5\},\{1,4,6\},\{1,5,6\},$$
$$\{2,3,6\},\{2,4,5\},\{2,5,6\},\{3,4,5\},\{3,4,6\}.$$

This example was given by the statistician F. Yates in 1936 in a paper which discussed the use of balanced designs in the construction of agricultural experiments. Many examples of such designs had been discussed by mathematicians over the previous hundred years, but his paper crystallised the idea and lead to much work on the subject by both statisticians and mathematicians.

Definition 9.1

A (v,k,λ) design is a collection of k-element subsets (called **blocks**) of a v-element set S, where $k < v$, such that each pair of elements of S occur together in exactly λ blocks. Such a design is also known as a **balanced incomplete block design** (BIBD).

The adjective "balanced" refers to the existence of λ, and "incomplete" refers to the requirement that $k < v$ (so that no block contains all the elements). We have already seen some examples of balanced designs.

Example 9.3

(i) League schedules. The games of a league schedule for $2n$ teams form a $(2n, 2, 1)$ design; if each team plays each other twice in a season, the games form a $(2n, 2, 2)$ design.

(ii) The seven point plane (Figure 9.1) is a $(7, 3, 1)$ design.

(iii) The design of Example 9.2 is a $(6, 3, 2)$ design.

(iv) An affine plane of order n, described at the end of Section 7.4, is a $(n^2, n, 1)$ design.

In particular, an affine plane of order 3 is a $(9, 3, 1)$ design. It consists of 12 blocks of size 3, such that every pair of elements occur together once in a block. Block designs with $k = 3$ and $\lambda = 1$ were among the first designs to be studied.

Definition 9.2

A $(v, 3, 1)$ design is called a **Steiner triple system** of order v and is often denoted by STS(v).

Steiner triple systems can exist only for certain values of v. To prove this, we first obtain the following general result.

Theorem 9.1

Suppose that a (v, k, λ) design has b blocks. Then each element occurs in precisely r blocks, where

$$\lambda(v - 1) = r(k - 1) \quad \text{and} \quad bk = vr. \tag{9.1}$$

Proof

Choose any element x, and suppose it occurs in r blocks. In each of these r blocks it makes a pair with $k - 1$ other elements; so altogether there are $r(k - 1)$ pairs in the blocks involving x. But x is paired with each of the $v - 1$ other elements λ times, so the number of pairs involving x is also $\lambda(v - 1)$. So $\lambda(v - 1) = r(k - 1)$. This shown that r is independent of the choice of x, since it is uniquely determined by v, k and λ.

To prove that $bk = vr$, note first that each block has k elements, and so the b blocks contain bk elements altogether (including repetitions). But each element x occurs r times in the blocks, so we must have $bk = vr$.

Example 9.4

In an affine plane of order n, i.e. a $(n^2, n, 1)$ design, we have $1.(n^2-1) = (n-1)r$, so $r = n + 1$. Also $bn = n^2r$, so $b = n(n + 1) = n^2 + n$, as in Section 7.4.

Example 9.5

No $(11, 6, 2)$ design can exist, since it would require $2(11 - 1) = 5r$, i.e. $r = 4$, and $6b = 44$, which is clearly impossible.

Theorem 9.2

A STS(v) can exist only if $v \equiv 1$ or $3 \pmod{6}$.

Proof

Suppose a $(v, 3, 1)$ design exists. Then $v-1 = 2r$ and $3b = vr$, so that $v = 2r+1$ (which is odd) and $b = \frac{1}{6}v(v - 1)$. If $v = 6u + 5$ then $b = \frac{1}{6}(6u + 5)(6u + 4)$ is not an integer; so we must have $v \equiv 1$ or $3 \pmod{6}$.

Note that $v = 7$ and 9, already dealt with, are of this form. Steiner systems are so named because Steiner discussed them in an 1853 paper, having come across them in a geometrical setting. But it had already been shown by Kirkman in 1847 that not only was the condition $v \equiv 1$ or $3 \pmod{6}$ necessary, but it was also sufficient. So STS(v) exists if and only if $v \equiv 1$ or $3 \pmod{6}$.

Example 9.6

The sets $\{1, 2, 5\}, \{2, 3, 6\}, \ldots, \{9, 10, 13\}, \{10, 11, 1\}, \ldots, \{13, 1, 4, \}$ and the sets $\{1, 3, 9\}, \{2, 4, 10\}, \ldots, \{5, 7, 13\}, \{6, 8, 1\}, \ldots, \{13, 2, 8\}$ form a STS(13). Note that the blocks are obtained from $\{1, 2, 5\}$ and $\{1, 3, 9\}$ by adding 1 successively to each element, and working modulo 13. This is similar to the STS(7) of Figure 9.1 which is obtained from the initial block $\{1, 2, 4\}$ and working modulo 7. **Why** this works will be explained in Section 9.5.

Further progress on designs is assisted by the representation of designs by matrices.

Definition 9.3

The **incidence matrix** of a (v, k, λ) design is the $b \times v$ matrix $A = (a_{ij})$ defined by

$$a_{ij} = \begin{cases} 1 & \text{if the } i\text{th block contains the } j\text{th element,} \\ 0 & \text{otherwise.} \end{cases}$$

For example, the incidence matrix of the seven-point plane of Figure 9.1, with blocks $\{1, 2, 4\}, \{2, 3, 5\}, \ldots, \{7, 1, 3\}$ is

$$
\begin{bmatrix}
1 & 1 & 0 & 1 & 0 & 0 & 0 \\
0 & 1 & 1 & 0 & 1 & 0 & 0 \\
0 & 0 & 1 & 1 & 0 & 1 & 0 \\
0 & 0 & 0 & 1 & 1 & 0 & 1 \\
1 & 0 & 0 & 0 & 1 & 1 & 0 \\
0 & 1 & 0 & 0 & 0 & 1 & 1 \\
1 & 0 & 1 & 0 & 0 & 0 & 1
\end{bmatrix}. \tag{9.2}
$$

The first row has 1 in positions $1, 2, 4$ since the first block is $\{1, 2, 4\}$. Note that the **rows** correspond to the **blocks** and the **columns** correspond to the **elements**. Note too that the matrix A depends on the order in which the blocks and the elements are taken. However, it turns out that the important properties of A do not depend on the particular orders chosen.

Theorem 9.3

If A is the incidence matrix of a (v, k, λ) design, then

$$
A^T A = (r - \lambda)I + \lambda J \tag{9.3}
$$

where r is as in (9.1), and where J is the $v \times v$ matrix with every entry equal to 1.

Proof

The (i, j)th entry of $A^T A$ is the scalar product of the ith row of A^T and the jth column of A, i.e. of the ith and jth columns of A. Thus the (i, i) diagonal entry is the scalar product of the ith column of A with itself, and so is the number of 1s in the ith column. But this is just the number of blocks containing the ith element, which is r.

If $i \neq j$, the scalar product of the ith and jth columns is just the number of places in which both columns have a 1. This is the number of blocks containing both the ith and jth elements, which is λ. So all diagonal entries of $A^T A$ are r, and all non-diagonal entries are λ.

An important consequence of this result is the fact that a (v, k, λ) design cannot contain fewer blocks than elements. This result was first obtained by the statistician R.A. Fisher in 1940.

Theorem 9.4

In any (v, k, λ) design, $b \geq v$.

Proof

We give a matrix proof, using Theorem 9.3 and basic properties of determinants (a purely combinatorial proof is outlined in Exercise 9.21).

Let A be the incidence matrix of the design. Then, denoting the determinant of a matrix M by $|M|$, we have

$$\begin{vmatrix} r & \lambda & \lambda & \dots & \lambda \\ \lambda & r & \lambda & \dots & \lambda \\ \lambda & \lambda & r & \dots & \lambda \\ \vdots & & & & \\ \lambda & \lambda & \lambda & \dots & r \end{vmatrix} = \begin{vmatrix} r & \lambda & \lambda & \dots & \lambda \\ \lambda-r & r-\lambda & 0 & & 0 \\ \lambda-r & 0 & r-\lambda & & 0 \\ \vdots & \vdots & \vdots & & \vdots \\ \lambda-r & 0 & 0 & \dots & r-\lambda \end{vmatrix}$$

on subtracting the first row from each of the other rows. We now add to the first column the sum of all the other columns to obtain

$$|A^T A| = \begin{vmatrix} r+(v-1)\lambda & \lambda & \lambda & \dots & \lambda \\ 0 & r-\lambda & 0 & \dots & 0 \\ 0 & 0 & r-\lambda & \dots & 0 \\ \vdots & & & & \\ 0 & 0 & 0 & \dots & r-\lambda \end{vmatrix}$$

$$= \{r+(v-1)\lambda\}(r-\lambda)^{v-1}$$

$$= rk(r-\lambda)^{v-1}$$

since, by (9.1), $r+(v-1)\lambda = r+r(k-1) = rk$. Now $k < v$ so, by (9.1), $r > \lambda$, and so $|A^T A| \neq 0$. Now $A^T A$ is a $v \times v$ matrix and so its rank $\rho(A^T A)$ must be v. But $\rho(A^T A) \leq \rho(A)$, where $\rho(A) \leq$ number of rows of A; so $\rho(A^T A) \leq b$. Thus $v \leq b$ as required.

Example 9.7

We can show that no $(25, 10, 3)$ design can exist. If it did, then (9.1) would give $72 = 9r$ and $10b = 25r$, whence $r = 8$ and $b = 20$. But this gives $b < v$.

A (v, k, λ) design with $b = v$ is called a **symmetric design**. Note that if $b = v$ it follows from (9.1) that $r = k$ and

$$\lambda(v-1) = k(k-1). \tag{9.4}$$

Also, (9.3) reduces to

$$A^T A = (k-\lambda)I + \lambda J. \tag{9.5}$$

A symmetric design is not so named because its incidence matrix is symmetric - usually it is not! - but because of a symmetry between some of the properties of the blocks and the elements. In a symmetric design we have, since $r = k$,

every block contains k elements;
every element is in k blocks.

We also have

$$\text{every pair of elements is in } \lambda \text{ blocks}$$

and we are now going to establish that

$$\text{every pair of blocks intersect in } \lambda \text{ elements.}$$

Theorem 9.5

If A is the incidence matrix of a symmetric design then $AA^T = A^T A$.

Proof

Since $AJ = kJ$ and $JA = rJ$, it follows from $r = k$ that $AJ = JA$. Thus A commutes with J, and so A commutes with $(k - \lambda)I + \lambda J = A^T A$. Thus $AA^T = A\{(k-\lambda)I + \lambda J\}A^{-1} = \{(k-\lambda)I + \lambda J\}AA^{-1} = (k-\lambda)I + \lambda J = A^T A$.

Corollary 9.6

In a symmetric (v, k, λ) design, every pair of blocks intersect in λ elements.

Proof

The (i, j)th entry of AA^T is the product of the ith row of A and the jth column of A^T, i.e. the product of the ith and jth rows of A. But this is just the number of columns in which the ith and jth rows both have 1, i.e. the number of elements in both the ith and jth blocks. By Theorem 9.5 this number is λ whenever $i \neq j$.

Once consequence of these symmetry properties is that if we take the incidence matrix A of a symmetric design and interchange rows and columns (to obtain A^T), the A^T is also the incidence matrix of a symmetric design, called the **dual** design. For example, the transpose of the incidence matrix of the seven-point plane of Figure 9.1 is

$$\begin{bmatrix} 1 & 0 & 0 & 0 & 1 & 0 & 1 \\ 1 & 1 & 0 & 0 & 0 & 1 & 0 \\ 0 & 1 & 1 & 0 & 0 & 0 & 1 \\ 1 & 0 & 1 & 1 & 0 & 0 & 0 \\ 0 & 1 & 0 & 1 & 1 & 0 & 0 \\ 0 & 0 & 1 & 0 & 1 & 1 & 0 \\ 0 & 0 & 0 & 1 & 0 & 1 & 1 \end{bmatrix}$$

which is the incidence matrix of the seven-point plane with blocks $\{1, 5, 7\}$, $\{1, 2, 6\}$, $\{2, 3, 7\}$, $\{1, 3, 4\}$, $\{2, 4, 5\}$, $\{3, 5, 6\}$, $\{4, 6, 7\}$. (Note however that the relabelling $1 \to 7, 2 \to 6, \ldots, 7 \to 1$ reveals that this second plane is in fact the original one in disguise!)

Complementary designs

Given a (v, k, λ) design D, we can obtain another design $\overline{D}$ from it in which the blocks are the complements of the blocks of D. $\overline{D}$ is called the **complementary design** of D. In Example 9.1, the blocks of size 4 form such a design which is complementary to the seven-point plane.

Theorem 9.7

Let D be a (v, k, λ) design on a set S, with blocks $B_1, \ldots, B_b$. Then the sets $\overline{B}_i = S \backslash B_i$ form a $(v, v - k, \lambda')$ design where $\lambda' = b - 2r + \lambda$, provided $\lambda' > 0$.

Proof

Since $|B_i| = k$ for each i, it is clear that $|\overline{B}_i| = v - k$ for each i. We have to show that every pair of elements of S lie in exactly λ' of the blocks $\overline{B}_i$. Now if $x, y \in S$, then $x, y \in \overline{B}_i$ precisely when neither x nor y belongs to B_i. But by the inclusion-exclusion principle, the number of blocks B_i containing neither x nor y is

$$b - \text{(no. of blocks containing } x) - \text{(no. of blocks containing } y)$$
$$+ \text{(no. of blocks containing both } x \text{ and } y)$$
$$= b - 2r + \lambda.$$

Example 9.8

The complementary design to the seven-point plane is a $(7, 4, 2)$ design since $\lambda' = b - 2r + \lambda = 7 - 6 + 1 = 2$. The blocks of this design are the groups of size 4 in Example 9.1.

Theorem 9.8

The complement of a symmetric design is also symmetric.

Proof

If D is symmetric, $b = v$, and so $\overline{D}$ also has $b = v$ blocks.

9.2 Resolvable Designs

Example 9.9 (The Kirkman schoolgirls problem)

In 1850, Kirkman posed the following problem: "Fifteen young ladies in a school walk out three abreast for seven days in succession: it is required to arrange them daily, so that no two shall walk twice abreast".

If such an arrangement exists, the triples of girls will form a STS(15). According to the proof of Theorem 9.2, any such design has $b = \frac{1}{6}.15.14 = 35$ blocks; the problem is to group these blocks into seven groups of 5 blocks, the blocks in each group forming a partition of the set of 15 schoolgirls. This requirement is similar to that for the games in a league schedule for $2n$ teams, where the games (pairs) are arranged into $2n - 1$ groups of n games, the pairs in each group forming a partition of the set of $2n$ teams.

Definition 9.4

A (v, k, λ) design on a set S is **resolvable** if the blocks can be arranged into r groups so that each group forms a partition of S. The groups are then called the **resolution** or **parallel classes**.

(Note why there have to be r groups: each element occurs in r blocks, and has to occur in precisely one block of each group. Note too that a resolvable design can only exist when $k|v$.)

Here, for example, is a solution to the Kirkman schoolgirls problem. In this resolvable $(15, 3, 1)$ design the blocks are grouped into seven groups of 5 blocks, each group partitioning $\{1, \ldots, 15\}$. Read the groups horizontally.

$1, 8, 15$	$2, 4, 10$	$3, 7, 12$	$5, 6, 9$	$11, 13, 14$
$2, 9, 15$	$3, 5, 11$	$4, 1, 13$	$6, 7, 10$	$12, 14, 8$
$3, 10, 15$	$4, 6, 12$	$5, 2, 14$	$7, 1, 11$	$13, 8, 9$
$4, 11, 15$	$5, 7, 13$	$6, 3, 8$	$1, 2, 12$	$14, 9, 10$
$5, 12, 15$	$6, 1, 14$	$7, 4, 9$	$2, 3, 13$	$8, 10, 11$
$6, 13, 15$	$7, 2, 8$	$1, 5, 10$	$3, 4, 14$	$9, 11, 12$
$7, 14, 15$	$1, 3, 9$	$2, 6, 11$	$4, 5, 8$	$10, 12, 13$

Affine planes

Affine planes were constructed in Section 7.4. Starting with a complete set of MOLS of order n, we constructed an affine plane with n^2 points and $n^2 + n$ lines. We also saw that the lines could be grouped into $n + 1$ groups of n parallel lines: in other words, the affine plane so obtained was **resolvable**. We now show that **all** $(n^2, n, 1)$ designs, no matter how they are obtained, must be resolvable.

Theorem 9.9

Every $(n^2, n, 1)$ design is resolvable.

Proof

From (9.1), we have $r = n + 1$ and $b = n^2 + n$. We first show that, given any block $B = \{b_1, \ldots, b_n\}$ and any element $x \notin B$, there is a **unique** block containing x which does not intersect B. (Compare this with the following:

given a line ℓ and a point P not on ℓ, there is a unique line through P which does not meet ℓ (i.e. which is parallel to ℓ).)

For each $b_i \in B$, there is a unique block B_i containing both b_i and x. Clearly $B_i \neq B_j$ whenever $i \neq j$, for otherwise b_i and b_j would be contained in both B and B_i, contradicting the fact that $\lambda = 1$. So we get n blocks B_i containing x and intersecting B. But altogether there are $r = n + 1$ blocks containing x; so there must be exactly one block containing x and disjoint from B.

Next we note that if C_1 and C_2 are any two blocks disjoint from B then C_1 and C_2 do not intersect. For suppose $z \in C_1 \cap C_2$; then z would be in more than one block which is disjoint from B, contrary to what we have just proved.

So consider the $n^2 - n$ elements not in B. For each such x, there is a unique block containing x and disjoint from B. Since each block has size n, we therefore have $\frac{n^2-n}{n} = n - 1$ blocks disjoint from B and disjoint from one another, and thus forming a resolution class with B. So each block is contained in a resolution class of n blocks, and so the design is resolvable, two blocks being in the same class if and only if they are disjoint.

We can now use resolvability to get back from affine planes to MOLS.

Theorem 9.10

An affine plane of order n exists $\Leftrightarrow$ a complete set of $n - 1$ MOLS of order n exists.

Proof

The proof of $\Leftarrow$ was given in Section 7.4, so we now consider the reverse implication. Suppose that a $(n^2, n, 1)$ design exists. It is necessarily resolvable, and it has $r = n + 1$ resolution classes. Select two such classes, say $\{B_1, \ldots, B_n\}$ and $\{C_1, \ldots, C_n\}$. Since each point of the plane lies in exactly one B_i and one C_j we can give it unique "coordinates" (i, j).

There remain $n - 1$ other resolution classes, and we construct a latin square for each. For the resolution class $\{E_1, \ldots, E_n\}$, define an $n \times n$ matrix $E = (e_{ij})$ by

$$e_{ij} = k \qquad \text{where the point } (i, j) \text{ lies in } E_k.$$

First we check that E is a latin square. If $e_{ij} = e_{iJ}$ with $j \neq J$ then (i, j) and (i, J) lie in E_k but also in B_i, contradicting $\lambda = 1$. So no row of E has repeated elements, and a similar argument applies to columns.

Finally we check that if $\{F_1, \ldots, F_n\}$ is another resolution class then E and F are orthogonal. So suppose $e_{ij} = e_{IJ}(= k)$ and $f_{ij} = f_{IJ}(= \ell)$. Then (i, j) and (I, J) both lie in E_k and also in F_ℓ, again contradicting $\lambda = 1$.

So we obtain $n - 1$ mutually orthogonal latin squares, as required.

Example 9.10

Take the affine plane constructed in Example 7.10. Take

$$
\begin{aligned}
B_1 &= \{1,2,3\}, & B_2 &= \{4,5,6\}, & B_3 &= \{7,8,9\}, \\
C_1 &= \{1,4,7\}, & C_2 &= \{2,5,8\}, & C_3 &= \{3,6,9\}, \\
E_1 &= \{1,6,8\}, & E_2 &= \{2,4,9\}, & E_3 &= \{3,5,7\}, \\
F_1 &= \{1,5,9\}, & F_2 &= \{2,6,7\}, & F_3 &= \{3,4,8\}.
\end{aligned}
$$

Then, for example, the point $(1,1)$ is 1, $(1,2)$ is 2, $(2,3)$ is 6. Since $1 \in E_1, e_{11} = 1$; since $2 \in E_2, e_{12} = 2$; since $6 \in E_1, e_{23} = 1$. In this way we obtain

$$
E = \begin{bmatrix} 1 & 2 & 3 \\ 2 & 3 & 1 \\ 3 & 1 & 2 \end{bmatrix} \quad \text{and} \quad F = \begin{bmatrix} 1 & 2 & 3 \\ 3 & 1 & 2 \\ 2 & 3 & 1 \end{bmatrix}.
$$

9.3 Finite Projective Planes

In searching for symmetric designs it is natural to look for such designs with $\lambda = 1$. The seven-point plane is one such design. In general, if we have a design with $k = n + 1$ and $\lambda = 1$ then $\lambda(v - 1) = k(k - 1)$ gives $v - 1 = n^2 + n$, so that the design is a $(n^2 + n + 1, n + 1, 1)$ design. Conversely, any $(n^2 + n + 1, n + 1, 1)$ design must, by (9.1), have $n^2 + n = rn$ and $b(n + 1) = (n^2 + n + 1)r$, so that $r = n + 1 = k$ and $b = n^2 + n + 1 = v$; so the design must be symmetric.

Definition 9.5

For $n \geq 2$, a **finite projective plane** (FPP) of order n is a $(n^2 + n + 1, n + 1, 1)$ design.

Thus in a FPP there are equal numbers of blocks and elements. They mimic the lines and points of a geometry: any two blocks (lines) intersect in one element (point), and any two points lie on a unique line. The seven-point plane is a FPP of order 2.

Example 9.11

The blocks $\{1,2,4,10\}, \{2,3,5,11\}, \{3,4,6,12\}, \{4,5,7,13\}, \{5,6,8,1\}, \ldots,$ $\{10,11,13,6\}, \{11,12,1,7\}, \{12,13,2,8\}, \{13,1,3,9\}$ are the lines of a FPP of order 3, i.e. the blocks form a $(13,4,1)$ design. Note the cyclic nature of the design (modulo 13). It is now known that such a cyclic FPP of order p exists for all primes p (modulo $p^2 + p + 1$).

We now show the fundamental connection between projective and affine planes. An artist does not draw parallel lines as parallel; the two sides of a road converge as the road disappears into the distance. The artist is using projective geometry to represent euclidean (affine) geometry. So let us follow this idea, and, for each of the $n + 1$ resolution classes of an affine plane of order n, add a "point at infinity" to each block of that class. Thus, for each $i \leq n + 1$, we add a new point ∞_i to each block of the ith class. The blocks now have size $n + 1$, there are $n^2 + n + 1$ points, and it should be clear that the property of any two points lying on a unique line is preserved, except that there is no line containing any two of the new points ∞_i. So if we introduce one new line $\{\infty_1, \ldots, \infty_{n+1}\}$ we have a $(n^2 + n + 1, n + 1, 1)$ design, i.e. a FPP of order n. The line thus introduced is sometimes called the **line at infinity**.

Example 9.12

Take the affine plane of order 3 used in Example 9.10, and introduce four "points at infinity", which we denote by $10, 11, 12, 13$. We obtain a FPP of order 3 with the following lines:

$$
\begin{array}{lll}
\{1, 2, 3, 10\} & \{4, 5, 6, 10\} & \{7, 8, 9, 10\} \\
\{1, 4, 7, 11\} & \{2, 5, 8, 11\} & \{3, 6, 9, 11\} \\
\{1, 6, 8, 12\} & \{2, 4, 9, 12\} & \{3, 5, 7, 12\} \\
\{1, 5, 9, 13\} & \{2, 6, 7, 13\} & \{3, 4, 8, 13\} \\
\{10, 11, 12, 13\}.
\end{array}
$$

We have thus established one half of the following theorem.

Theorem 9.11

There exists a FPP of order $n \Leftrightarrow$ there exists an affine plane of order n.

Proof

We have dealt with $\Leftarrow$ above. So now suppose there exists a FPP of order n. Take any line $\ell = \{p_1, \ldots, p_{n+1}\}$ of the plane. (We are going to treat it as if it were the line at infinity.) Since ℓ intersects every other line in exactly one point, each other line contains exactly one of the p_i. So if we throw away ℓ and remove its points wherever they occur, we end up with $n^2 + n$ lines each with n points. Any two points lie in exactly one line, namely the remnant of the line of the FPP on which they lay. So we have an affine plane of order n, in which, for each i, the remnants of the lines of the FPP containing p_i form a resolution class.

Combining Theorems 9.10 and 9.11 together, we obtain the following remarkable result.

Theorem 9.12

The following statements are equivalent.

(i) There exists a complete set of $n - 1$ MOLS of order n.

(ii) There exists an affine plane of order n.

(iii) There exists a finite projective plane of order n.

In view of Theorem 7.2 and the note following, FPPs of prime power order exist. It is conjectured that if n is not a prime power then no FPP of order n exists. This has been confirmed for $n = 6$ and 10, but the case $n = 12$ is still open.

9.4 Hadamard Matrices and Designs

We begin this section by introducing a family of matrices all of whose entries are ± 1. From these matrices an important family of designs will be obtained.

Two vectors will be called **orthogonal** if their scalar product is 0. For example, any two row vectors in the matrix

$$\begin{bmatrix} 1 & 1 & 1 & 1 \\ 1 & -1 & 1 & -1 \\ 1 & 1 & -1 & -1 \\ 1 & -1 & -1 & 1 \end{bmatrix} \tag{9.6}$$

are orthogonal, as are any two columns.

Definition 9.6

A $(+1, -1)$ matrix is a matrix all of whose entries are ± 1. An $n \times n$ $(+1, -1)$ matrix H is a **Hadamard matrix of order** n if $HH^T = H^T H = nI$.

Note that $HH^T = nI$ is essentially saying that H is invertible with inverse $\frac{1}{n}H^T$. Since a matrix commutes with its inverse, either of the properties $HH^T = nI$ and $H^T H = nI$ follows from the other. Note too that $HH^T = nI$ is precisely the property that any two rows of H are orthogonal, and similarly $H^T H = nI$ is equivalent to demanding that the columns of H are orthogonal. Hadamard matrices are so named after Jacques Hadamard who, in 1893, showed that any real $n \times n$ matrix H, whose entries h_{ij} all satisfy $|h_{ij}| \leq 1$, has determinant at most $n^{n/2}$, equality occurring only if $HH^T = nI$. But since then Hadamard matrices have cropped up in many areas of combinatorics, and have even been behind the sending of photographs from Mars to Earth. This will be explained in the final section.

There is a straightforward way of constructing Hadamard matrices of order 2^m.

Let

$$H_0 = [1], \qquad H_1 = \begin{bmatrix} 1 & 1 \\ 1 & -1 \end{bmatrix}$$

and, for each $m \geq 1$, define H_m inductively by

$$H_m = \begin{bmatrix} H_{m-1} & H_{m-1} \\ H_{m-1} & -H_{m-1} \end{bmatrix}.$$

Then it is straightforward to check that H_m is a Hadamard matrix of order 2^m; consider the scalar product of any two of its rows. Thus, for example, H_2 is just the matrix (9.6).

But for other values of n does a Hadamard matrix of order n exist?

We shall show that if $n > 2$ then n has to be a multiple of 4. Before proving this, it is helpful to note that if we have a Hadamard matrix and we multiply any row or column by -1 then the resulting matrix is still Hadamard; orthogonality of the rows and columns is preserved. So we can always **normalise** a Hadamard matrix, i.e. make every entry in the first row and the first column equal to $+1$.

Theorem 9.13

If there exists a Hadamard matrix H of order $n > 2$, then n must be a multiple of 4.

Proof

Assume that H has been normalised, so that the first row has $+1$ in each position. Since the rows are orthogonal, the second row must have the same number of $+1$s and -1s; so there must be $\frac{n}{2}$ $+1$s and $\frac{n}{2}$ -1s, so that n is necessarily even. By rearranging the order of the columns, we can suppose that H has first two rows

$$\begin{matrix} 1 & 1 & \cdots & 1 & 1 & 1 & \cdots & 1 \\ 1 & 1 & \cdots & 1 & -1 & -1 & \cdots & -1. \end{matrix}$$

If $n > 2$, consider the third row of H. Suppose that h of its first $\frac{n}{2}$ entries are $+1$, and k of its second $\frac{n}{2}$ entries are $+1$. Then $\frac{n}{2} - h$ of its first $\frac{n}{2}$ entries are -1, and $\frac{n}{2} - k$ of its second $\frac{n}{2}$ entries are -1. Since the first and third rows are orthogonal,

$$h - (\frac{n}{2} - h) + k - (\frac{n}{2} - k) = 0,$$

i.e. $h + k = \frac{n}{2}$. Also, since the second and third rows are orthogonal,

$$h - (\frac{n}{2} - h) - k + (\frac{n}{2} - k) = 0,$$

i.e. $h = k$. Thus $h = k = \frac{n}{4}$, and n has to be a multiple of 4.

Since a similar argument can be applied to the columns, we have:

Corollary 9.14

In any normalised Hadamard matrix of order $4m$, any two columns other than the first have $+1$s together in exactly m places.

Hadamard designs

If a Hadamard matrix of order $4m$ exists, we can obtain a design from it. The basic idea is to take a normalised matrix, remove its first row and column, replace each -1 by 0, and interpret the resulting matrix as the incidence matrix of a design.

Theorem 9.15

If a Hadamard matrix of order $4m$ exists, then a $(4m - 1, 2m - 1, m - 1)$ design exists.

Proof

Let H be a normalised Hadamard matrix of order $4m$. As in the proof of Theorem 9.13, each row and column of H, apart from the first, must have $2m$ $+1$s and $2m$ -1s. Remove the first row and column of H, and replace each -1 in the resulting matrix by 0. This results in a $(4m-1) \times (4m-1)$ $(0,1)$-matrix A in which every row and every column has $2m$ 0s and $2m - 1$ 1s.

We interpret A as the incidence matrix of a (necessarily symmetric) design. It has $4m - 1$ blocks and $4m - 1$ elements; each block contains $2m - 1$ elements and each element lies in $2m - 1$ blocks. Consider now any two elements. The columns of A corresponding to these two elements have (by Corollary 9.14) $m - 1$ 1s together, i.e. these two elements occur together in exactly $m - 1$ blocks. So the design is balanced with $\lambda = m - 1$.

Definition 9.7

A $(4m - 1, 2m - 1, m - 1)$ design is called a **Hadamard design**.

Example 9.13

The doubling construction gives, from (9.6), a normalised Hadamard matrix of order 8. Removing the first row and column and changing every -1 to 0 gives the following matrix:

$$
\begin{bmatrix}
0 & 1 & 0 & 1 & 0 & 1 & 0 \\
1 & 0 & 0 & 1 & 1 & 0 & 0 \\
0 & 0 & 1 & 1 & 0 & 0 & 1 \\
1 & 1 & 1 & 0 & 0 & 0 & 0 \\
0 & 1 & 0 & 0 & 1 & 0 & 1 \\
1 & 0 & 0 & 0 & 0 & 1 & 1 \\
0 & 0 & 1 & 0 & 1 & 1 & 0
\end{bmatrix}.
\tag{9.7}
$$

This is the incidence matrix of a $(7, 3, 1)$ design, i.e. a FPP of order 2, whose blocks are:

$$\{2,4,6\}, \{1,4,5\}, \{3,4,7\}, \{1,2,3\}, \{2,5,7\}, \{1,6,7\}, \{3,5,6\}.$$

There is, in fact, essentially only one FPP of order 2. The plane just found can be "converted" into the plane of Figure 9.1 by replacing $1, 2, 3, 4, 5, 6, 7$ by $1, 6, 5, 4, 2, 3, 7$ respectively.

The construction presented in the proof of Theorem 9.15 can be reversed: given a Hadamard design, replace each 0 in the incidence matrix by -1 and add a first row and column with all entries $+1$. The resulting matrix is Hadamard. So we have the following result.

Theorem 9.16

A Hadamard matrix of order $4m$ exists $\Leftrightarrow$ a Hadamard $(4m-1, 2m-1, m-1)$ design exists.

This result enables us to construct Hadamard matrices of order $4m$ provided that we can construct the corresponding design. Many methods of constructing Hadamard designs have been found, but we give only the simplest of these constructions.

Let p be any prime of the form $p = 4m - 1$, and take the **squares** of $1, 2, \ldots, \frac{1}{2}(p-1) \pmod{p}$. For example, if $p = 7$, the squares of $1, 2, 3$ are $1, 4, 2 \pmod 7$ since $9 \equiv 2 \pmod 7$. So we obtain the set $\{1, 2, 4\}$ which we observe is precisely the set used to generate the seven-point plane in Example 9.1. In general, starting with any prime $p = 4m - 1$, the set of squares of $1, 2, \ldots, \frac{1}{2}(p-1) \pmod p$ acts as a starting block for a cyclic Hadamard $(4m - 1, 2m - 1, m - 1)$ design. This will be justified in the next section (Theorem 9.18). See the appendix for details of squares $\pmod p$.

Example 9.14

Take $p = 11$. The squares of $1, \ldots, 5 \pmod{11}$ are $1, 4, 9, 5, 3$. So take the set $\{1, 3, 4, 5, 9\}$ as the starting block and obtain other blocks by successively adding $1 \pmod{11}$ to each entry of the block. The resulting $(11, 5, 2)$ design has incidence matrix

$$
\begin{bmatrix}
1 & 0 & 1 & 1 & 1 & 0 & 0 & 0 & 1 & 0 & 0 \\
0 & 1 & 0 & 1 & 1 & 1 & 0 & 0 & 0 & 1 & 0 \\
0 & 0 & 1 & 0 & 1 & 1 & 1 & 0 & 0 & 0 & 1 \\
1 & 0 & 0 & 1 & 0 & 1 & 1 & 1 & 0 & 0 & 0 \\
0 & 1 & 0 & 0 & 1 & 0 & 1 & 1 & 1 & 0 & 0 \\
0 & 0 & 1 & 0 & 0 & 1 & 0 & 1 & 1 & 1 & 0 \\
0 & 0 & 0 & 1 & 0 & 0 & 1 & 0 & 1 & 1 & 1 \\
1 & 0 & 0 & 0 & 1 & 0 & 0 & 1 & 0 & 1 & 1 \\
1 & 1 & 0 & 0 & 0 & 1 & 0 & 0 & 1 & 0 & 1 \\
1 & 1 & 1 & 0 & 0 & 0 & 1 & 0 & 0 & 1 & 0 \\
0 & 1 & 1 & 1 & 0 & 0 & 0 & 1 & 0 & 0 & 1
\end{bmatrix} \tag{9.8}
$$

and from this we obtain the following Hadamard matrix of order 12 (where $+$, $-$ stand for $+1, -1$ respectively)

$$
\begin{bmatrix}
+ & + & + & + & + & + & + & + & + & + & + & + \\
+ & + & - & + & + & + & - & - & - & + & - & - \\
+ & - & + & - & + & + & + & - & - & - & + & - \\
+ & - & - & + & - & + & + & + & - & - & - & + \\
+ & + & - & - & + & - & + & + & + & - & - & - \\
+ & - & + & - & - & + & - & + & + & + & - & - \\
+ & - & - & + & - & - & + & - & + & + & + & - \\
+ & - & - & - & + & - & - & + & - & + & + & + \\
+ & + & - & - & - & + & - & - & + & - & + & + \\
+ & + & + & - & - & - & + & - & - & + & - & + \\
+ & + & + & + & - & - & - & + & - & - & + & - \\
+ & - & + & + & + & - & - & - & + & - & - & +
\end{bmatrix}
.
$$

It is conjectured that a Hadamard matrix of order $4n$ exists for all positive integers n. But this is far from being proved.

9.5 Difference Methods

The seven-point plane can be constructed by starting with $\{1, 2, 4\}$ and obtaining further blocks by adding $1 \pmod 7$ successively, to obtain $\{2, 3, 5\}, \{3, 4, 5\}$, etc. What is special about the choice of $1, 2, 4$ which makes the method work? Similarly, the block $\{1, 3, 4, 5, 9\}$ was used in Example 9.14 to obtain a Hadamard design; what is special about this choice?

Consider the **differences** between elements of $\{1, 2, 4\}$ modulo 7. They are $\pm(2 - 1), \pm(4 - 2), \pm(4 - 1)$, i.e. $\pm 1, \pm 2, \pm 3$, i.e. $1, 2, 3, 4, 5, 6$, i.e. all the non-zero numbers $\pmod 7$, each occurring **once**. Similarly, consider the differences between elements of $\{1, 3, 4, 5, 9\} \pmod{11}$; they are $\pm 2, \pm 3, \pm 4, \pm 8, \pm 1, \pm 2, \pm 6$, $\pm 1, \pm 5, \pm 4$, i.e. $\pm 1, \pm 2, \pm 3, \pm 4, \pm 5$ twice, i.e. all the non-zero numbers $\pmod{11}$ each occurring twice. Note that the first gives rise to a design with $\lambda = 1$, and the second gives rise to a design with $\lambda = 2$.

Definition 9.8

(i) Let $\mathbb{Z}_v$ denote the integers modulo v. A k-element subset $D = \{d_1, \ldots, d_k\}$ of $\mathbb{Z}_v$ is called a cyclic (v, k, λ) **difference set** if $\lambda > 0, 2 \leq k < v$, and every non-zero $d \in \mathbb{Z}_v$ can be expressed as $d = d_i - d_j$ in exactly λ ways.

(ii) If D is a difference set, the set $D + a = \{d_1 + a, \ldots, d_k + a\}$ is called a **translate** of D.

Thus the seven-point plane of Example 9.1 consists of the translates of $\{1, 2, 4\}$. This is a special case of the following general result.

Theorem 9.17

If $D = \{d_1, \ldots, d_k\}$ is a cyclic (v, k, λ) difference set then the translates $D + i, 0 \leq i \leq v - 1$, are the blocks of a symmetric (v, k, λ) design.

Proof

Here $D + i = \{d_1 + i, \ldots, d_k + i\}$. Clearly there are v translates, each of size k. So we have only to check the balance property. Two elements x, y are in the same translate $D + a$ if and only if $x = a + d_i, y = a + d_j$ for some $i \neq j$, i.e. $x - a = d_i, y - a = d_j$, i.e. $(x - a, y - a)$ is one of the λ pairs (d_i, d_j) such that $d_i - d_j = x - y$.

We can now see why the "starter" block $\{1, 2, 4, 10\}$ in Example 9.12 gives rise to a FPP of order 3. The differences are $\pm 1, \pm 3, \pm 9, \pm 2, \pm 8, \pm 6$, i.e. all the non-zero numbers (mod 13), each once, so $\{1, 2, 4, 10\}$ is a $(13, 4, 1)$ difference set.

Example 9.15

(i) $\{1, 2, 5, 15, 17\}$ is a $(21, 5, 1)$ difference set in $\mathbb{Z}_{21}$, and its translates form a FPP of order 4.

(ii) $\{1, 2, 7, 19, 23, 30\}$ is a $(31, 6, 1)$ difference set, leading to a FPP of order 5.

(iii) $\{1, 3, 4, 5, 9\}$ is a $(11, 5, 2)$ difference set and its translates form a $(11, 5, 2)$ design as in Example 9.14.

(iv) The set $\{2, 6, 7, 8, 10, 11\}$, which is the complement of the difference set of (iii), is itself a $(11, 6, 3)$ difference set, the translates of which are the blocks of a $(11, 6, 3)$ design. This design is the complementary design (Theorem 9.7) of the Hadamard $(11, 5, 2)$ design of Example 9.14.

It is now known that a $(p^2 + p + 1, p + 1, 1)$ difference set exists for all primes p, leading to a cyclic FPP in each case. It cannot be emphasised too strongly how useful this difference set technique is in the construction of symmetric designs. Of course the method is useful only if difference sets can be constructed. One method of construction was mentioned in the previous section in connection with Hadamard designs, so we now show why that method of construction works.

The number-theoretic facts required for the proof can be found in the appendix.

Theorem 9.18

Let $p = 4m - 1$ be prime. Then the non-zero squares in $\mathbb{Z}_p$ form a $(p, \frac{1}{2}(p - 1), \frac{1}{4}(p - 3))$ difference set.

Proof

Since $(-x)^2 = x^2$, each square arises twice; as shown in the appendix, exactly half of the non-zero elements of $\mathbb{Z}_p$ are squares. Thus there are $\frac{1}{2}(p-1)$ non-zero squares.

Let w be any non-zero square, say $w \equiv t^2 (\mathrm{mod}\, p)$. Then for each representation $1 = x^2 - y^2$ as a difference of squares, there corresponds the representation $w = (tx)^2 - (ty)^2$. Conversely, if s is the inverse of $t(\mathrm{mod}\, p)$, then corresponding to each representation $w = x^2 - y^2$ we have a representation $1 = s^2 t^2 = s^2 w = (sx)^2 - (sy)^2$. So all non-zero squares w in $\mathbb{Z}_p$ have the same number of representations as a difference of squares.

Further, since $p \equiv 3(\mathrm{mod}\, 4)$, -1 is not a square in $\mathbb{Z}_p$ and the non-squares are precisely the negatives of the squares. So, corresponding to any representation $z = x^2 - y^2$ of the non-square z, we have the representation $-z = y^2 - x^2$ of the square $-z$. So all non-squares and squares have the same number, say λ, of representations as a difference of squares. The value of λ is obtained from $\lambda(v-1) = k(k-1)$: we have $\lambda(p-1) = \frac{1}{2}(p-1) \cdot \frac{1}{2}(p-3)$, whence $\lambda = \frac{1}{4}(p-3)$.

This difference method can be extended to non-symmetric designs. For example consider the construction of a league schedule for $2n$ teams as described in Section 8.1. Apart from the game involving ∞, the first round games were

$$1 \,\mathrm{v}\, (2n-2), 2 \,\mathrm{v}\, (2n-3), \ldots, (n-1) \,\mathrm{v}\, n.$$

The pairs $\{1, 2n - 2\}, \{2, 2n - 3\}, \ldots, \{n - 1, n\}$ have differences $\pm(2n - 3), \pm(2n - 5), \ldots, \pm 1$, i.e. every non-zero member of $\mathbb{Z}_{2n-1}$.

Again, consider the STS of order 13 presented in Example 9.7. The blocks are the translates of the two initial blocks $\{1, 2, 5\}$ and $\{1, 3, 9\}$. These two blocks have differences $\pm 1, \pm 3, \pm 4$ and $\pm 2, \pm 6, \pm 8$, i.e. all the non-zero members of $\mathbb{Z}_{13}$, each exactly once.

Example 9.16

The blocks $\{1, 2, 9\}, \{1, 3, 17\}, \{1, 5, 14\}$ have differences ± 1, ± 7, ± 8, ± 2, ± 14, ± 16, ± 4, ± 9, ± 13 (mod 19), i.e. each non-zero element of $\mathbb{Z}_{19}$ once. So the translates will form a STS(19).

9.6 Hadamard Matrices and Codes

A binary code is a collection of n-digit binary sequences, called **codewords**. If codewords are transmitted then it is possible that errors will arise due to interference, and so the received codewords may differ in some places from those that were sent. The basic idea behind an error-correcting code is to choose the codewords to be sufficiently different from each other so that even if some errors in transmission occur, each received word is "closer" to the transmitted codeword than to any other. We have here the concept of the **distance** between two codewords, namely the number of places in which they differ. If all the codewords are chosen so that each pair differ in at least $2t + 1$ places, then, even if t errors are made in the transmission of a codeword, the received binary sequence will still be closer to the original than to any other, and hence can be correctly decoded as the codeword nearest to it.

Definition 9.9

A binary code of length n is a set C of n-digit binary sequences, called **codewords**. The (Hamming) **distance** $d(\mathbf{x}, \mathbf{y})$ between any two codewords $\mathbf{x}, \mathbf{y}$ is the number of places in which they differ. If $d(\mathbf{x}, \mathbf{y}) \geq 2t + 1$ for all codewords $\mathbf{x}, \mathbf{y}, \mathbf{x} \neq \mathbf{y}$, the code C is said to be a t-**error-correcting** code.

Example 9.17

The four codewords $0000000, 1111111, 1010101, 0101010$ differ from each other in at least 3 places, and so form a 1-error-correcting code. For example, if 1010101 is sent, and due to interference 1110101 is received, 1110101 is closer to 1010101 than to any other codeword and hence will be decoded correctly.

There are two conflicting aspects of a code. Given n, it would be desirable to have the minimum distance between any two codewords as big as possible (to enhance error correction) but it would also be desirable to have as many codewords as possible. But these properties conflict: you cannot have too many codewords which are all a large distance apart. So we have a fundamental problem: given n and k, how many binary sequences of length n can we find such that each pair of binary sequences differ in at least k places?

We shall look at a particular case of this problem, namely when $k = \lceil \frac{n}{2} \rceil$. First we consider the case when n is odd.

Lemma 9.19

Suppose that there are N binary sequences of length $n = 2m - 1$, any two of which differ in at least m places. Then $N \leq n + 1$.

Proof

Consider the N sequences as the rows of an $N \times n$ $(0, 1)$-matrix. Let S denote the sum of all the distances $d(\mathbf{x}, \mathbf{y})$ between the sequences:

$$S = \sum_{\mathbf{x} \neq \mathbf{y}} d(\mathbf{x}, \mathbf{y}).$$

Here the sum is over all $\binom{N}{2}$ pairs of distinct sequences $\mathbf{x}, \mathbf{y}$.

We shall count S in two different ways. For each pair $\mathbf{x}, \mathbf{y}, d(\mathbf{x}, \mathbf{y}) \geq m$; so we obtain immediately that

$$S \geq \binom{N}{2} m. \tag{9.9}$$

Now consider the jth column of the matrix. If it contains a_j 0s and b_j 1s, then $a_j + b_j = N$ and, since each of the a_j 0s gives a difference with each of the b_j 1s, there is a contribution of $a_j b_j$ to S from this column. So $S = \sum_{j=1}^{n} a_j b_j$. Now it is easily checked (e.g. by calculus) that if $x + y = N$, then the maximum possible value of xy is $\frac{N^2}{4}$. So each $a_j b_j$ is at most $\frac{N^2}{4}$, and we have:

$$S \leq n \cdot \frac{N^2}{4}. \tag{9.10}$$

From (9.9) and (9.10) we obtain

$$m \frac{N(N-1)}{2} \leq \frac{nN^2}{4},$$

whence

$$(2m - n)N \leq 2m,$$

i.e. (since $n = 2m - 1$)

$$N \leq n + 1.$$

Example 9.18

How many codewords of length 11 can be found, such that each pair differ in at least 6 places?

Solution

By Lemma 9.19, we cannot hope to find more than 12 such codewords. But, further, we can find as many as 12 by using the Hadamard design of Example 9.14. The 11 rows of the incidence matrix (9.8) each have 5 1s; but any two rows share only two 1s; so any two rows differ in $2 \times (5 - 2) = 6$ places. So if we take the 11 rows as codewords, along with the row of all 1s (which will differ from the other 11 in 6 places), we obtain 12 codewords as required.

Corollary 9.20

Let $\mathcal{C}$ be a code of length $n = 2m$, containing N codewords, each pair of which differ in at least m places. Then $N \leq 2n$. Further, if a Hadamard matrix of order n exists, then there exists such a code with $2n$ codewords.

Proof

Consider the codewords of $\mathcal{C}$ which begin with 0. Omitting this 0 from them, we obtain codewords of length $2m - 1$ differing in at least m places. By Lemma

9.19 there can be at most $2m$ such codewords. Similarly, there can be at most $2m$ codewords of C beginning with 1; thus C has at most $4m = 2n$ codewords.

If $n = 2$, then $C = \{11, 10, 01, 00\}$ satisfies the requirements. So suppose now that there exists a Hadamard matrix H of order $n > 2$. Then $n = 4u$ for some u. Any two rows of H differ in exactly $2u = \frac{n}{2}$ places, and agree in exactly $\frac{n}{2}$ places.

Let A denote the matrix obtained from H by replacing -1 by 0, and let $\overline{A}$ denote the matrix obtained from A by interchanging 0s and 1s. Then any two rows of A differ in at least $\frac{n}{2}$ places, as do any two rows of $\overline{A}$ and a row of A and a row of $\overline{A}$ differ in either $\frac{n}{2}$ or all n places. So the rows of A and $\overline{A}$ give the required $2n$ codewords.

Example 9.19

Take the Hadamard matrix H_5 constructed by the duplication method of Section 9.4. It has 32 rows and columns. Let A denote the matrix obtained from H_5 by replacing each -1 by 0, and let $\overline{A}$ be obtained from A by interchanging 0s and 1s. Then the rows of A and $\overline{A}$ form a code of 64 codewords of length 32, each differing in at least 16 places; so the code is 7-error-correcting. Such a code was used in the 1972 Mars Mariner 9 space probe to Mars, to send photographs back to Earth. Each photograph consisted of lots of dots of different shades of grey (64 shades needed 6-digit binary sequences to represent them, since $2^6 = 64$), and the sequence of coded shades was encoded using the 7-error-correcting code which we have just described. The resulting pictures were remarkably good!

More recent spaces probes have used much more sophisticated codes, as do compact discs and other modern gadgets. We refer the reader to Hill [12] for a good introduction to coding theory.

There is one final connection between codes and Hadamard designs with which we shall bring this chapter to a close. Before describing it, we need first to find a bound on the number of codewords in a t-error-correcting code.

Theorem 9.21

If C is a binary t-error-correcting code of length n, then

$$|C| \cdot \left\{ \binom{n}{0} + \binom{n}{1} + \cdots + \binom{n}{t} \right\} \leq 2^n. \tag{9.11}$$

Proof

Any two codewords of C differ in at least $2t + 1$ places. Thus any sequence differing from a codeword $\mathbf{x}$ in at most t places will be corrected to $\mathbf{x}$.

Now, for each i, $0 \leq i \leq t$, there are $\binom{n}{i}$ binary sequences of length n differing from $\mathbf{x}$ in i places; so the number of sequences correctable to $\mathbf{x}$ is $\sum_{i=0}^{t} \binom{n}{i}$. Thus, since there are $|C|$ possible choices of $\mathbf{x}$, there must be at least $|C| \sum_{i=0}^{t} \binom{n}{i}$

binary sequences of length n. Since there are precisely 2^n binary sequences of length n, the result follows.

A code C is said to be **perfect** if there is equality in (9.11). For such a code C, **every** binary sequence is correctable to a codeword of C, i.e. every sequence is at distance at most t from a unique codeword. Note that for equality to hold in (9.11), we must have $\binom{n}{0} + \cdots + \binom{n}{t}$ equal to a power of 2.

Example 9.20

A perfect 1-error-correcting code C of length n can exist only when

$$|C| \cdot (1 + n) = 2^n.$$

Consider the case $n = 7$, so that $|C| = 16$. We can construct such a code by taking the seven rows of the matrix (9.7), the seven codewords obtained from these by interchanging 0s and 1s, and the row of all 0s and the row of all 1s. This gives 16 codewords as follows.

0	1	0	1	0	1	0		1	0	1	0	1	0	1
1	0	0	1	1	0	0		0	1	1	0	0	1	1
0	0	1	1	0	0	1		1	1	0	0	1	1	0
1	1	1	0	0	0	0		0	0	0	1	1	1	1
0	1	0	0	1	0	1		1	0	1	1	0	1	0
1	0	0	0	0	1	1		0	1	1	1	1	0	0
0	0	1	0	1	1	0		1	1	0	1	0	0	1
0	0	0	0	0	0	0		1	1	1	1	1	1	1

Another way of obtaining these codewords is to take four of them which are linearly independent over $\mathbb{Z}_2$, and to take all linear combinations of them over $\mathbb{Z}_2$. In other words, we choose four linearly independent codewords $c_1, \ldots, c_4$ and take all linear combinations $\lambda_1 c_1 + \lambda_2 c_2 + \lambda_3 c_3 + \lambda_4 c_4$ where each λ_i is 0 or 1. For example, if we take the first four codewords above,

$$c_1 = 0101010, \quad c_2 = 1001100, \quad c_3 = 0011001, \quad c_4 = 1110000,$$

we then have

$$0100101 \;=\; c_2 + c_3 + c_4,$$

$$1010101 \;=\; c_2 + c_3,$$

etc. Thus C consists of all linear combinations (mod 2) of the rows of the matrix

$$\mathcal{G} = \begin{bmatrix} 0 & 1 & 0 & 1 & 0 & 1 & 0 \\ 1 & 0 & 0 & 1 & 1 & 0 & 0 \\ 0 & 0 & 1 & 1 & 0 & 0 & 1 \\ 1 & 1 & 1 & 0 & 0 & 0 & 0 \end{bmatrix}$$

which is called a **generator matrix** for C. C is called a **linear code**. Linear codes have special advantages over other codes. They all include **0** as a codeword, and the minimum distance between any two codewords is just the minimum distance of codewords from $\mathbf{0}$, i.e. the minimum number of 1s in a non-zero codeword (see Exercise 9.19). In the above example, this number is clearly 3, so the code is 1-error-correcting. For each n of the form $2^m - 1$ there is a perfect linear 1-error-correcting code of length n whose generating matrix has k rows, where $k = n - m$. These **Hamming codes** are described in [**12**].

For $t > 1$, perfect t-error correcting codes are rare. But there is one jewel corresponding to the identity

$$\binom{23}{0} + \binom{23}{1} + \binom{23}{2} + \binom{23}{3} = 2^{11} = 2^{23-12}.$$

A perfect 3-error-correcting code of length 23 was presented by Golay in 1949, and our final task is to explain what this code is.

Example 9.21 (The perfect Golay code $\mathcal{G}_{23}$)

Recall from Example 9.15(iv) that $\{2, 6, 7, 8, 10, 11\}$ is a $(11, 6, 3)$ difference set (mod 11) whose translates form the design complementary to a Hadamard $(11, 5, 2)$ design. Let A denote the incidence matrix of this $(11, 6, 3)$ design:

$$A = \begin{bmatrix} 0 & 1 & 0 & 0 & 0 & 1 & 1 & 1 & 0 & 1 & 1 \\ 1 & 0 & 1 & 0 & 0 & 0 & 1 & 1 & 1 & 0 & 1 \\ 1 & 1 & 0 & 1 & 0 & 0 & 0 & 1 & 1 & 1 & 0 \\ \vdots & & & & & & & & & & \vdots \end{bmatrix}.$$

Then define the matrix

$$B = \begin{bmatrix} 1 & & & 0 & & \\ 1 & & I_{11} & 0 & & A \\ \vdots & & & \vdots & & \\ 1 & & & 0 & & \\ \hline 0 & 0 & \cdots & 0 & 1 & 1 & \cdots\cdots & 1 \end{bmatrix}.$$

Then B is a 12×24 $(0, 1)$-matrix. The code $\mathcal{G}_{23}$ is then obtained by removing the first column of B and taking all linear combinations of the rows of the resulting 12×23 matrix. The reason for first including the first column of B is that it makes the arithmetic of the argument simpler. It is clear that, since the rows of A differ in 6 places, any two rows of B differ in at least 8 places. Our aim is to show that **any** two linear combinations of rows of B differ in at least 8 places, and so any two codewords of $\mathcal{G}_{23}$ will differ in at least 7 places. We have to show that, if $\mathbf{x}$ is a non-zero linear combination of rows of B, then $w(\mathbf{x}) \geq 8$ where $w(\mathbf{x})$, the **weight** of $\mathbf{x}$, is the number of 1s in $\mathbf{x}$. We do this by a sequence of observations. Let $\mathcal{V}$ denote the set of linear combinations of rows of B.

(i) $w(\mathbf{x} + \mathbf{y}) = w(\mathbf{x}) + w(\mathbf{y}) - 2\mathbf{x} \cdot \mathbf{y}$ for all $\mathbf{x}, \mathbf{y} \in \mathcal{V}$.

This follows since $w(\mathbf{x}) + w(\mathbf{y})$ counts twice any 1s in the same place in $\mathbf{x}$ and $\mathbf{y}$, but these 1s add to 0 (mod 2) and hence have to be subtracted twice.

(ii) Every row of B has weight 8 or 12.

(iii) If $\mathbf{x}$ and $\mathbf{y}$ are any two rows of B, $\mathbf{x} \cdot \mathbf{y}$ is even.

This follows since any two rows of A have 1s together in exactly 3 places.

(iv) If $\mathbf{x}$ and $\mathbf{y}$ are any two rows of $B, w(\mathbf{x} + \mathbf{y})$ is a multiple of 4.

This follows from (i), (ii) and (iii).

(v) If $\mathbf{x} \in \mathcal{V}$ and $\mathbf{y}$ is a row of B, then $\mathbf{x} \cdot \mathbf{y}$ is even.

This is proved by induction. For the induction step consider $\mathbf{x} = \mathbf{z} + \mathbf{r}$ where $\mathbf{r}$ is a row of B and $\mathbf{z}$ is the sum of k rows of B. Then $\mathbf{x} \cdot \mathbf{y} = (\mathbf{z} + \mathbf{r}) \cdot \mathbf{y} \equiv \mathbf{z} \cdot \mathbf{y} + \mathbf{r} \cdot \mathbf{y}$ (mod 2), which is even since $\mathbf{z} \cdot \mathbf{y}$ is even by induction hypothesis and $\mathbf{r} \cdot \mathbf{y}$ is even by (iii).

(vi) If $\mathbf{x} \in \mathcal{V}$ then $w(\mathbf{x})$ is a multiple of 4.

Again, this follows by induction. For the induction step take $\mathbf{x} = \mathbf{z} + \mathbf{r}$ as in (v). Then $w(\mathbf{x}) = w(\mathbf{z}) + w(\mathbf{r}) - 2\mathbf{z} \cdot \mathbf{r}$. But $w(\mathbf{z})$ is a multiple of 4 by hypothesis, $w(\mathbf{r})$ is a multiple of 4 by (ii), and $\mathbf{z} \cdot \mathbf{r}$ is even by (v).

Since each $\mathbf{x} \in \mathcal{V}$ has weight a multiple of 4, it follows that, to prove $w(\mathbf{x}) \geq 8$, we need only show that $w(\mathbf{x}) = 4$ is impossible. This is achieved by considering the left half and the right half of each $\mathbf{x} \in \mathcal{V}$. Denote by $w_{\mathrm{L}}(\mathbf{x})$ and $w_{\mathrm{R}}(\mathbf{x})$ the weights of the left and right halves of $\mathbf{x}$ respectively.

(vii) $w_{\mathrm{L}}(\mathbf{x})$ is even for all $\mathbf{x} \in \mathcal{V}$. This follows since if $\mathbf{x}$ is the sum of an even number of rows of B then the 1s in the first column sum to $0 \pmod 2$, leaving an even number of 1s from I_{11}, while if $\mathbf{x}$ is the sum of an odd number of rows, $\mathbf{x}$ has 1 in the first position and an odd number of 1s from I_{11}.

(viii) $w_{\mathrm{L}}(\mathbf{x}) = 0, w_{\mathrm{R}}(\mathbf{x}) = 4$ is impossible.

For if $w_{\mathrm{L}}(\mathbf{x}) = 0$ then $\mathbf{x}$ is either $\mathbf{0}$ (in which case $w_{\mathrm{R}}(\mathbf{x}) = 0$) or the last row of B (in which case $w_{\mathrm{R}}(\mathbf{x}) = 12$).

(ix) $w_{\mathrm{L}}(\mathbf{x}) = w_{\mathrm{R}}(\mathbf{x}) = 2$ is impossible.

For if $w_{\mathrm{L}}(\mathbf{x}) = 2$ then $\mathbf{x}$ must be the sum of one or two of the first 11 rows of B, possibly together with the last row of B. But the weight of the sum of one or two rows of A is $6 > 2$, and if the last row of B is added the resulting $\mathbf{x}$ has $w_{\mathrm{R}}(\mathbf{x}) = 6 > 2$.

(x) $w_{\mathrm{L}}(\mathbf{x}) = 4, w_{\mathrm{R}}(\mathbf{x}) = 0$ is impossible.

Here $\mathbf{x}$ must be the sum of 3 or 4 of the first 11 rows of B. If $\mathbf{x}$ is the sum of 3 rows, let $\mathbf{r}$ be any other of the first 11 rows. Then, since $w_R(\mathbf{x}) = 0, w_R(\mathbf{x}+\mathbf{r}) = 6$, and $w_L(\mathbf{x}+\mathbf{r}) = 4$, giving $w(\mathbf{x}+\mathbf{r}) = 10$, contrary to (vi). If $\mathbf{x}$ is the sum of 4 rows of B, let t be one of them. Then $\mathbf{x} = \mathbf{z}+\mathbf{t}$ where $\mathbf{z}$ is the sum of 3 rows of B. Then $w_L(\mathbf{z}) = 4$, and $w_R(\mathbf{z}) = w_R(\mathbf{x}+\mathbf{t}) = w_R(\mathbf{t}) = 6$ since $w_R(\mathbf{x}) = 0$. So $w(\mathbf{z}) = 10$, again contradicting (vi). So $w_L(\mathbf{x}) = 4, w_R(\mathbf{x}) = 0$ is impossible.

Thus the minimum weight of non-zero sums $\mathbf{x} \in \mathcal{V}$ is indeed 8, and the minimum weight of a non-zero codeword of $\mathcal{G}_{23}$ is indeed 7. Thus $\mathcal{G}_{23}$ is a 3-error-correcting code as required.

The code $\mathcal{G}_{23}$ has remarkable connections with Steiner systems; the reader is referred to [2] for further details.

Exercises

Exercise 9.1

Show that no (a) $(17, 9, 2)$, (b) $(21, 6, 1)$ design exists.

Exercise 9.2

How would you construct a $(13, 9, 6)$ design?

Exercise 9.3

Use (9.1) to show that (a) $vr(k-1)\lambda = r^2(k-1)^2 + r(k-1)\lambda$, (b) $(k-1)\lambda = (k-1)r - (v-k)\lambda$.

Exercise 9.4

Show that, in a symmetric (v, k, λ) design, $k - 1 < \sqrt{\lambda v} < k$.

Exercise 9.5

Deduce from the proof of Theorem 9.3 that if a (v, k, λ) symmetric design exists then its incidence matrix A satisfies $|A|^2 = k^2(k-\lambda)^{v-1}$. Hence show that if a symmetric design exists with v even then $k - \lambda$ must be a perfect square. Hence show that no $(34, 12, 4)$ or $(46, 10, 2)$ design can exist.

Exercise 9.6

Write down the parameters (b, v, r, k, λ) for the complementary design of (a) an affine plane of order n, (b) a $(4m - 1, 2m - 1, m - 1)$ Hadamard design.

Exercise 9.7

How would you construct a $(15, 7, 3)$ design?

Exercise 9.8

How would you construct a Hadamard matrix of order 24?

Exercise 9.9

Show that if $D = \{d_1, \ldots, d_k\}$ is a (v, k, λ) difference set in $\mathbb{Z}_k$ then $-D = \{-d_1, \ldots, -d_k\}$ and all translates of D are also (v, k, λ) difference sets.

Exercise 9.10

Verify that $\{1, 2, 3, 5, 6, 9, 11\}$ yields a $(15, 7, 3)$ design.

Exercise 9.11

Verify that $\{1, 2, 8, 20, 24, 45, 48, 50\}$ is a difference set (mod 57) which yields a FPP of order 7.

Exercise 9.12

Verify that $\{1, 8, 11, 12, 24\}$ and $\{1, 6, 15, 21, 23\}$ yield a $(41, 5, 1)$ design.

Exercise 9.13

Verify that the translates of $\{1, 2, 13\}, \{1, 4, 9\}, \{1, 3, 10\}, \{1, 5, 11\}$ form a STS(25).

Exercise 9.14

Verify that the sets $\{1,2,7\}, \{1,3,14\}, \{1,5,13\}, \{1,4,11\}$ have as differences all non-zero elements of $\mathbb{Z}_{27}$ except 9 and 18. Deduce that their translates, along with nine translates of $\{1,10,19\}$, form a STS(27).

Exercise 9.15

Take a 6×6 latin square L, and the 6×6 "natural" array N with $1,\ldots,36$ in natural order (cf. the 4×4 N at the beginning of Section 7.4). Call i,j **associates** if they are in the same row or column of N or if they are in positions in N where the corresponding entries in L are equal. Let $B_i = \{j \le 36 : i \text{ and } j \text{ are associates}\}$. Show that $B_1,\ldots,B_{36}$ are the blocks of a symmetric $(36,15,6)$ design.

Exercise 9.16

(a) Show that if A is a square $(0,1)$ matrix and B is obtained from A by replacing each 0 by -1 then $B = 2A - J$.
(b) Show that if A is the incidence matrix of a symmetric (v,k,λ) design then $B = 2A - J$ is a Hadamard matrix if and only if $v = 4(k - \lambda)$.
(c) Deduce that Exercise 9.15 enables a Hadamard matrix of order 36 to be constructed.

Exercise 9.17

Can you construct a Hadamard matrix of order $4m$ for all m up to 12?

Exercise 9.18

By following the proof of Lemma 9.19, show that if a code has N codewords of length n, each differing in at least d places, where $d > \frac{n}{2}$, then $N \le \frac{2d}{2d-n}$. (This is called Plotkin's bound.)

Exercise 9.19

Show that in a linear code the minimum distance between two non-equal codewords is the minimum weight of a non-zero codeword.

Exercise 9.20

Show that if a perfect 1-error-correcting code of length n exists then a STS(n) exists.

Exercise 9.21

(A non-matrix proof of Fisher's inequality.) Choose any block B, and let x_i denote the number of blocks intersecting B in exactly i elements. Show first that (a) $\sum_i x_i \binom{i}{2} = \binom{k}{2}(\lambda - 1)$, (b) $\sum_i i x_i = k(r - 1)$.

Let m denote the average value of $|B \cap C|$ taken over all blocks $C \neq B$. Thus (c) $m(b - 1) = \sum_i i x_i$.
Show that $\sum_i (i - m)^2 x_i \geq 0$ yields
(d) $(b - 1)k(k - 1)(\lambda - 1) + (b - 1)k(r - 1) \geq k^2(r - 1)^2$.

Now use (9.1) and Exercise 9.3 to deduce that $(r - k)(r - \lambda)(v - k) \geq 0$. This implies $r \geq k$, whence $b \geq v$.

(By now you will be glad of the matrix proof!)

Appendix

Arithmetic modulo n

Let $n \geq 2$ be an integer. If a, b are integers, we say that a is **congruent** to b modulo n (and we write $a \equiv b \pmod{n}$) whenever a and b differ by a multiple of n, i.e. whenever n divides $a - b$. Thus, for example,

$$8 \equiv 3 \,(\text{mod } 5), \quad 2 \equiv 10 \,(\text{mod } 4), \quad 20\,001 \equiv -99 \,(\text{mod } 100).$$

Let $\mathbb{Z}_n$ denote the set $\{0, 1, \ldots, n-1\}$ in which addition and multiplication are carried out $\text{mod } n$. (Sometimes 0 is replaced by n.) For example, in $\mathbb{Z}_9$, $7 \times 8 = 2$ since $56 \equiv 2 \,(\text{mod } 9)$.

Here are the addition and multiplication tables for $\mathbb{Z}_5$.

+	0	1	2	3	4
0	0	1	2	3	4
1	1	2	3	4	0
2	2	3	4	0	1
3	3	4	0	1	2
4	4	0	1	2	3

×	0	1	2	3	4
0	0	0	0	0	0
1	0	1	2	3	4
2	0	2	4	1	3
3	0	3	1	4	2
4	0	4	3	2	1

If p is prime, $\mathbb{Z}_p$ has the properties that characterise a field. Note that if $p|ab$ then $p|a$ or $p|b$, i.e. if $ab \equiv 0 \,(\text{mod } p)$, then $a \equiv 0$ or $b \equiv 0$. Thus the multiplication table for $\mathbb{Z}_p - \{0\}$ has no 0s. (This is unlike the situation in $\mathbb{Z}_6$, for example, where $2 \times 3 = 0$.) Next observe that if $p \nmid t$ (i.e. p does not divide t) then $t, 2t, \ldots, (p-1)t$ are all distinct $(\text{mod } p)$: for if $ta \equiv tb \,(\text{mod } p)$ then $p|t(a-b)$ whence $p|(a-b)$, i.e. $a \equiv b \,(\text{mod } p)$. Thus $t, 2t, \ldots, (p-1)t$ are just the same as $1, 2, \ldots, p-1$, but in a different order. The first consequence of this is that there must be some $s, 1 \leq s \leq p-1$, such that $st \equiv 1 \,(\text{mod } p)$. Thus each nonzero member t of $\mathbb{Z}_p$ has an **inverse** which we can denote by t^{-1}. For example, since $2 \times 3 \equiv 1$ in $\mathbb{Z}_5$, the inverse of 2 is 3, i.e. $2^{-1} = 3$ in $\mathbb{Z}_5$.

The second consequence arises from the congruence

$$t.2t \ldots (p-1)t \equiv 1.2. \ldots .(p-1) \quad (\text{mod } p)$$

i.e.

$$t^{p-1}(p-1)! \equiv (p-1)! \,(\text{mod } p).$$

Multiplying by the inverse of $(p-1)!$ gives **Fermat's theorem**

$$t^{p-1} \equiv 1 \,(\text{mod } p) \quad \text{whenever } p \nmid t.$$

For example,
$$3^{5-1} = 3^4 = 81 \equiv 1 \,(\text{mod } 5).$$

There is a further consequence of the existence of inverses. First note that the only two numbers $(\text{mod } p)$ which coincide with their inverse are 1 and -1. (For $x^2 \equiv 1 \Leftrightarrow (x-1)(x+1) = 0 \Leftrightarrow x \equiv 1$ or $-1 \,(\text{mod } p)$.). Consider $(p-1)! = 1.2.3 \ldots (p-2).(p-1)$. The numbers from $2, \ldots, p-2$ must consist of $\frac{1}{2}(p-3)$ pairs of numbers and their inverses; and so the product of all these numbers must be 1. So $(p-1)! \equiv 1.1.(p-1) \equiv -1 \,(\text{mod } p)$. Thus we obtain **Wilson's theorem**:
$$(p-1)! \equiv -1 \,(\text{mod } p).$$

Squares and non-squares in $\mathbb{Z}_p$

Let p be an odd prime: $p = 2k+1$. Then the numbers $1^2, 2^2, \ldots, k^2$ are all distinct $(\text{mod } p)$; for if $x^2 \equiv y^2$ then $(x-y)(x+y) \equiv 0 \,(\text{mod } p)$ so that $x \equiv y$ or $x \equiv -y \,(\text{mod } p)$. But $x \equiv -y$ is impossible, so $x \equiv y$.

Since $(p-x)^2 \equiv (-x)^2 \equiv x^2 \,(\text{mod } p)$, it follows that, of the $p-1 = 2k$ nonzero numbers $(\text{mod } p)$, exactly half are squares and half are non-squares. Note that:

(A1) if x and y are squares in $\mathbb{Z}_p$ then so is xy (for if $x \equiv u^2$ and $y \equiv v^2$ then $xy \equiv (uv)^2$);

(A2) if x is a square and y is a non-square in $\mathbb{Z}_p$ then xy is a non-square (for if $x = u^2$ and $xy = w^2$ then $y = (u^{-1}w)^2$ would be a square).

Further, note that if -1 is a square then $-1 \equiv x^2 \,(\text{mod } p)$ for some $x \in \mathbb{Z}_p$. Raising both sides to the power $\frac{1}{2}(p-1) = k$ gives $(-1)^k \equiv x^{p-1}$. But $x^{p-1} \equiv 1$ by Fermat's theorem, so $(-1)^k \equiv 1 (\text{mod } p)$. Thus k must be even and so $p \equiv 1 \,(\text{mod } 4)$. Thus:

(A3) if $p \equiv 3 \,(\text{mod } 4)$, -1 is a non-square.

It then follows from (A2) that:

(A4) if $p \equiv 3 \,(\text{mod } 4)$ then x is a square $\Leftrightarrow -x$ is a non-square.

Example. Take $p = 11$. The squares $(\text{mod } 11)$ are $1, 4, 9, 16 \equiv 5$ and $25 \equiv 3$. The negatives of $1, 3, 4, 5, 9$ are $10, 8, 7, 6, 2$, i.e. the non-squares.

Now consider $p \equiv 1 \,(\text{mod } 4)$; here we show that -1 is a square $(\text{mod } p)$.

Note that, if $p = 4k + 1$,

$$
\begin{aligned}
(p-1)! &= 1.2.\ldots.2k.(2k+1)\ldots 4k \\
&\equiv 1.2.\ldots.2k.(-2k)\ldots(-1) \pmod{p} \\
&\equiv (-1)^{2k}(2k)!(2k)! \equiv ((2k)!)^2,
\end{aligned}
$$

So that $(p-1)!$ is a square. But, by Wilson's theorem, $(p-1)! \equiv -1 \pmod{p}$, so -1 is a square. Thus we have:

(A5) if $p \equiv 1 \pmod 4$ then -1 is a square $\pmod p$, and x is a square $\Leftrightarrow -x$ is a square.

Example. Take $p = 13$. The squares (mod 13) are $1, 4, 9, 16 \equiv 3, 25 \equiv 12 \equiv -1, 36 \equiv 10$, and their negatives are $12, 9, 4, 10, 1, 3$, i.e. the squares.

Further number theoretic results can be found in the recent book by Jones and Jones [13].

Solutions

Chapter 1

1.1 9!

1.2 (a) $\binom{15}{4}\binom{12}{4}$, (b) $\binom{27}{8} - \binom{15}{8} - \binom{15}{7}\binom{12}{1}$, (c) $\sum_{r=5}^{8}\binom{15}{r}\binom{12}{8-r}$.

1.3 Yes, since $52! > 10^{66}$.

1.4 (a) $\binom{43}{3}\binom{6}{3}/\binom{49}{6} = 0.1765\ldots$.

    (b) $\binom{43}{2}\binom{6}{4}/\binom{49}{6} = 0.00097\ldots$.

    (c) $\binom{43}{1}\binom{6}{5}/\binom{49}{6} = 0.00002\ldots$.

1.5 1 in (a) 6724520, (b) 43949268

1.6 1 in $14\binom{34}{5} = 3895584$; so about 4 times as likely.

1.7 $\binom{10}{5}/2^5$.

1.9 Add (i) and (ii) of Theorem 1.7.

1.10 Equate coefficients of x^n on both sides. Then put $n = s + 1$.

1.11 (b) In resulting sum, terms cancel in pairs. (c) $\binom{n-1}{k}$ = coefficient of x^k in $(1+x)^n(\sum_{s=0}^{\infty}(-1)^s x^s) = \sum_{r+s=k}\binom{n}{r}(-1)^s = (-1)^k\sum_{r=0}^{k}(-1)^r\binom{n}{r}$.

1.12 LHS $= n\sum_k\binom{m}{k}\binom{n-1}{k-1} = n\binom{m+n-1}{m-1}$ by Exercise 1.10.

1.13 (a) $\binom{15+4-1}{15} = \binom{18}{3}$. (b) $\binom{14}{3}$ on putting $x = 1+u$, etc. (c) Put $x = 3+a, y = -1+b, z = 1+c, w = -2+d$ to get $\binom{17}{3}$.

1.14 Solve $x_1 + \ldots + x_4 + x_5 = 6$ with each $x_i \geq 0 : \binom{10}{4}$.

1.15 Consider a choice of 6 from 49 as a binary sequence of length 49 with exactly six 0s. Want no two 0s adjacent. By Example 1.17, number is $\binom{44}{6} > \frac{1}{2}\binom{49}{6}$.

1.16 (a) $10 \times 9 \times 8 \times 7$, (b) $\binom{10}{4}$, (c) 10^4, (d) $\binom{10+4-1}{4}$.

1.17 $\sum_{r=0}^{n} r\binom{n}{r} = n\sum_{1}^{n-1}\binom{n-1}{r-1} = n2^{n-1}$ by Theorem 1.7(i). Average size is
$\frac{n}{2} = \frac{1}{2^n}\sum_r r\binom{n}{r}$.

1.18 (a) Summing gives $1^2 + \ldots + n^2 = \binom{n+2}{3} + \binom{n+1}{3} = \frac{1}{6}n(n+1)(2n+1)$.
(b) Summing gives $1^2 + \ldots + n^2 = \sum_{r=1}^{n}\binom{r}{2} + \sum_{r=1}^{n}\binom{r+1}{2} = \binom{n+1}{3} + \binom{n+2}{3}$
by Theorem 1.8.

1.19 Let $S = \binom{n}{0} + \binom{n}{3} + \ldots$. Then de Moivre gives $e^{\frac{1}{3}n\pi i} = (1+w)^n = S + w\{\binom{n}{1} + \binom{n}{4} + \cdots\} + w^2\{\binom{n}{2} + \binom{n}{5} + \cdots\}$. Taking real parts gives
$\cos\frac{n\pi}{3} = S - \frac{1}{2}\{\binom{n}{1} + \binom{n}{2} + \binom{n}{4} + \binom{n}{5} + \cdots\} = S - \frac{1}{2}(2^n - S)$.

Chapter 2

2.1 (a) $a_n = 1 + \frac{1}{2} + \ldots + \frac{1}{2^{n-2}} + \frac{1}{2^{n-1}}a_1 = 1 + \frac{1}{2} + \ldots + \frac{1}{2^{n-1}} = 2(1 - \frac{1}{2^n})$,

(b) $3^n - 2^{n-1}$,

(c) $(2n-1)3^{n-1}$,

(d) $a_n = 4a_{n-1} - 3a_{n-2}$ has solution $a_n = A + B3^n$. For particular solution
try $a_n = K2^n$. Get $K = -4$. So try $a_n = A + B3^n - 2^{n+2}$. Get
$a_n = 3^{n+1} - 2^{n+2}$.

2.2 $b_n = F_{n+1}$.

2.3 Let c_n = number ending in 1 = number ending in 2. Then $d_n = d_{n-1} + 2c_n$.
But $c_n = d_{n-1} - c_{n-1}$, so get $d_n = 2d_{n-1} + d_{n-2}$. Since $d_1 = 3$ and
$d_2 = 7, d_n = \frac{1}{2}(1+\sqrt{2})^{n+1} + \frac{1}{2}(1-\sqrt{2})^{n+1}$.

2.4 (a) $f(x) = x + (\frac{1}{2}a_1 + 1)x^2 + (\frac{1}{2}a_2 + 1)x^3 + \cdots = \frac{1}{2}xf(x) + x(1 + x + x^2 + \cdots)$,
so $f(x)(1 - \frac{1}{2}x) = \frac{x}{1-x}$, i.e. $f(x) = \frac{x}{(1-x)(1-\frac{1}{2}x)} = 2(\frac{1}{1-x} - \frac{1}{1-\frac{1}{2}x})$ whence
$a_n = 2(1 - \frac{1}{2^n})$.

(b) $f(x) = (6x-1)(\frac{1}{1-3x} - \frac{1}{1-2x})$ whence $a_n = 6(3^{n-1} - 2^{n-1}) - (3^n - 2^n)$.

2.5 $f_n = f_{n-1} + f_{n-2}$, so $f_n = F_{n-1}$ $(n \geq 2)$.

2.6 $a_n = 3(2^{n-1} - 2^{n-2} + \cdots + (-1)^{n-2}2) = 6(-1)^n(1 - 2 + \cdots + (-2)^{n-2}) = 2^n + 2(-1)^n$.

2.7 $L_n = (\frac{1+\sqrt{5}}{2})^n + (\frac{1-\sqrt{5}}{2})^n$.

2.10 $f(x) = F_1 x + F_2 x^2 + (F_1 + F_2)x^3 + (F_2 + F_3)x^4 + \cdots = F_1 x + F_2 x^2 + x^2(F_1 x + \cdots) + x(F_2 x^2 + \cdots) = x + 2x^2 + x^2 f(x) + x(f(x) - x)$, whence
$f(x)(1 - x - x^2) = x + x^2$. Thus $f(x) = \frac{1}{1-x-x^2} - 1 = \frac{1}{\sqrt{5}}(\frac{\alpha}{1-\alpha x} - \frac{\beta}{1-\beta x}) - 1$
whence $F_n = \frac{1}{\sqrt{5}}(\alpha^{n+1} + \beta^{a+1})$.

2.11 (a) By induction. (b) $F_n F_{n+2} F_{n+1}^2 = \det M^{n+2} = (\det M)^{n+2} = (-1)^{n+2} = (-1)^n$. (c) Write all three matrices as in (a) and then equate top left entries.

2.12 Induction step: $F_1 + \cdots + F_k + F_{k+1} = (F_{k+2} - 2) + F_{k+1} = F_{k+3} - 2$.

2.13 (a) $F_{2n} - 1$, (b) $F_{2n+1} - 1$, (c) $(-1)^{n-1}F_{n-1}$.

2.14 a_n = no. in which 1 stays put + no. in which 1 and 2 change places
    = $a_{n-1} + a_{n-2} + 1$. So $a_n = F_n - 1$.

2.15 (a) There are g_{n-1} such subsets of $\{1, \ldots, n\}$ not containing n, and g_{n-2}
    containing n; so $g_n = g_{n-1} + g_{n-2}$. (b) Take $p = k, q = n - k$. (c) $F_{n+1} = \sum_k \binom{n+1-k}{k}$.

2.17 Label points $1, \ldots, 2n$. If 1 is joined to $2r + 2$, circle is divided into two
    parts, one with $2r$ vertices $2, \ldots, 2r + 1$ and the other with $2(n - r - 1)$
    vertices. So $a_n = \sum_r a_r a_{n-r-1}$. Check that $a_1 = 1, a_2 = 2$.

2.18 Use (2.14).

2.19 Induction step: $d_{k+1} = kd_k + kd_{k-1} > kd_k > k.(k - 1)! = k!$.

2.20 Number of comparisons is $\leq 1 + 2 + \cdots + (n - 1) = \frac{1}{2}n(n - 1)$: same as
    bubblesort.

2.21 Eliminate y_n from two given equations. Auxiliary equation has roots 1 and
    $\frac{4}{5}$, so $x_n = A + (\frac{4}{5})^n B$, so $x_n \to A$. But $x_0 = A + B$ and $A + \frac{4}{5}B = x_1 = \frac{3}{5}x_0 + \frac{1}{2}y_0$. Eliminate B to get $A = \frac{5}{2}y_0 - x_0$.

2.22 d_n.

Chapter 3

3.1 Use Corollary 3.2.

3.2 $p = 3n + 3, 2q = 4n + 2 + 2n + 2 = 6n + 4$, so $q = p - 1$. Apply Theorem
    3.4(iii).

3.3 Each component must have $\geq 1 + \frac{1}{2}(p - 1) > \frac{p}{2}$ vertices, so there can be
    only one component.

3.4 (a) 4, (b) 40.

3.5 $q - (p - 1)$.

3.6 (a) Spanning tree has 4 edges, so one of the x_i must have 2 edges from it.

   (b) The x_i joined to both a and b can be chosen in 3 ways; there are then
       2 choices of edge from each of the other x_i: so $3 \times 2^2 = 12$.

   (c) 100×2^{99}.

3.7 (a) Choose in order AE, DC, AC, AB. (b) Choose AE, AC, DC, AB.

3.8 Using Prim, choose HM, HEK, HA, GEK.

3.9 11 flights required. $4 + 4\sqrt{2} + 3\sqrt{5}$.

3.10 (a) Suppose planar; $p = 6, q = 9$ gives $r = 5$. Bipartite, so get $18 = 2q \geq 4r = 20$, contradiction.

(b) Yes - draw 2 edges "outside".

(c) No. If so, $p = 8, q = 19, r = 13$, so $38 = 2q \geq 3r = 39$.

(d) No. If so, $p = 11, q = 10, r = 11$, so $40 = 2q \geq 4r = 44$.

3.11 6.

3.12 (a) (i) No. (ii) Yes (draw edges $y_i x_{i+1}$ "outside").

(b) a_n = no. using $x_n y_n$ + no. not using $x_n y_n = a_{n-1} + 2a_{n-2}$. So $a_n = \frac{1}{3}(2^{n+2} - (-1)^n)$.

3.13 (a) $p - q + r = 2$ and $2q \geq 3r$, so $2q \geq 3(2 + q - p)$, i.e. $q \leq 3p - 6$. For $K_5, 3p - 6 = 9 < 10 = q$.

(b) $p - q + r = 2$ and $2q \geq gr$, so $2q \geq g(2 - p + q)$, i.e. $(g - 2)q \leq g(p - 2)$.

(c) Use $g = 4$ for $K_{3,3}$ and $g = 5$ for Petersen graph.

3.14 Follow Example 3.14. $2q = 3p, r = s + h, 2q = 4s + 6h, p - q + r = 2$ yield $s = 6$.

3.15 The lines of cut and circular arcs form a graph with $2n$ vertices of degree 3 on outside and $\binom{n}{2}$ vertices of degree 4 inside. So $p = 2n + \binom{n}{2}, 2q = 6n + 4\binom{n}{2}$, so $r = n + \binom{n}{2} + 2$, including infinite region.

3.16 If g_n denotes the number containing edge $x_0 x_n$, then $h_n = 2h_{n-1} + g_{n-1}$ and $g_n = h_n - h_{n-1}$; thus $h_n = 3h_{n-1} - h_{n-2}$.

Chapter 4

4.1 (a) B and W must alternate. (b) $m = |B| = |W| = n$ if hamiltonian.

4.2 (a) (i) and (ii) are hamiltonian. (iii) is not by Exercise 4.1(a).

(b) None.

(c) (iii).

4.3 All hamiltonian; only octahedron is eulerian.

4.4 Only (a) is planar.

4.5 $00000 - 01000 - 11000 - 10000 - 10100 - 11100 - 01100 - 00100 - 00110 - 01110 - 11110 - 10110 - 10010 - 11010 - 01010 - 00010 - 00011 - 01011 - 11011 - 10011 - 10111 - 11111 - 01111 - 00111 - 00101 - 01101 - 11101 - 10101 - 10001 - 11001 - 01001 - 00001 - 00000$.

4.6 Let $A = \{j : v_i$ and v_j are adjacent$\}$ and $B = \{j : v_{j-1}$ and v_p are adjacent}. Then $A \subseteq \{2, \ldots, p-1\}, |A| \geq \frac{p}{2}, B \subseteq \{3, \ldots, p\}, |B| \geq \frac{p}{2}$; so $A \cap B \neq \emptyset$.

4.7 (a) Imitate proof of Dirac. $|A| = \deg(v_1)$, $|B| = \deg(v_p)$. Since $\deg(v_1) + \deg(v_p) \geq p$, get $A \cap B \neq \emptyset$ again.

   (b) If there are non-adjacent vertices u, w such that $\deg(u) + \deg(w) \leq p-1$, then $\mathcal{G}$ has at most $\binom{p-2}{2} + (p-1) = \frac{1}{2}(p-1)(p-2) + 1$ edges.

   (c) Take a vertex v in K_p and remove all but one edge from v.

4.8 Remove A: $(2+4) + 13 = 19$. Remove B: $(6+6) + 9 = 21$. Upper bound: the spanning tree AE, EC, CD, AB give upper bound $23(AECDBA)$.

4.9 Exact value is 37.

4.11 (i) Take vertices $0, 1, 2$, and all possible directed edges. Eulerian circuit gives 001122021 in cyclic order.

   (ii) Take vertices $00, 01, \ldots, 22$, and draw directed edge from ij to jl. One solution is $000101112122210220120200211$.

4.12 K_7, with a loop at each vertex, is eulerian. With vertices $0, \ldots, n, K_{n+1}$ with loops is eulerian $\Leftrightarrow$ n even. So arrangement is possible $\Leftrightarrow$ n even. For $n = 6$, one eulerian circuit is $0011223344556602461350362514 0$, giving cyclic arrangement of dominoes $(0,0), (0,1), (1,1), (1,2), \ldots$ etc.

4.13 K_n eulerian $\Leftrightarrow$ n odd. Required arrangement corresponds to eulerian circuit. For n even, duplicate each edge to get an arrangement with every pair adjacent twice.

Chapter 5

5.1 $\frac{16!}{(4!)^5}$.

5.2 $\frac{30!}{(7!)^2(8!)^2(2!)^2}$.

5.3 $\binom{26}{12}\frac{12!}{2^6 6!}$.

5.4 $\frac{8!}{2^3 3!} \times 2$.

5.5 (a) One of the parts must have two elements: choose it in $\binom{n}{2}$ ways.

   (b) There is either one set with 3 elements or two sets with two. In second case choose 4 elements and then partition into two pairs. Alternatively use (5.2) and iteration.

5.6 Induction step: $S(k+1, 3) = S(k,2) + 3S(k,3) > 3S(k,3) > 3 \cdot 3^{k-2} = 3^{(k+1)-2}$.

5.7 n is in a part with ℓ others, $0 \leq \ell \leq n-k$; so $S(n,k) = \sum_{\ell=0}^{n-k} \binom{n-1}{\ell} S(n-$

$1 - \ell, k - 1)$. Put $m = n - 1 - \ell$.

$$\begin{aligned} B(n) &= 1 + \sum_{k=2}^{n} S(n,k) = 1 + \sum_{k=2}^{n} \sum_{m=k-1}^{n-1} \binom{n-1}{m} S(m, k-1) \\ &= 1 + \sum_{m=1}^{n-1} \binom{n-1}{m} \sum_{k=2}^{m+1} S(m, k-1) \\ &= 1 + \sum_{m=1}^{n-1} \binom{n-1}{m} B(m). \end{aligned}$$

5.8 Use Theorem 5.6. $B(10) = 115\,975$.

5.9 $$\begin{aligned} B(k+1) &= \sum_{m=0}^{k} \binom{k}{m} B(m) &&= \tfrac{1}{e} \sum_{m=0}^{k} \binom{k}{m} \sum_{j=0}^{\infty} \tfrac{j^m}{j!} \\ &= \tfrac{1}{e} \sum_{j=0}^{\infty} \tfrac{1}{j!} \sum_{m=0}^{k} \binom{k}{m} j^m \\ &= \tfrac{1}{e} \sum_{j=0}^{\infty} \tfrac{1}{j!} (1+j)^k &&= \tfrac{1}{e} \sum_{j=0}^{\infty} \tfrac{(j+1)^{k+1}}{(j+1)!}. \end{aligned}$$

5.10 n either forms a 1-cycle on its own or can be slotted into a cycle in one of the $s(n, k-1)$ permutations of $1, \ldots, n-1$ (think of n inserted after one of the other $n-1$ elements). $s(5,2) = 50, s(6,2) = 274$.

5.11 $\chi = 3, 3, 2$. $\chi' = 4, 5, 4$.

5.12 Each colour can be used at most $\alpha(G)$ times.

5.13 (a) Vertices $1, \ldots, 8$ get colours $1, 2, 3, 4, 3, 1, 2, 3$.

(b) Vertices $8, \ldots, 1$ get colours $1, 1, 2, 3, 4, 2, 5, 1$.

5.14 (a) Vertices $8, 1, 3, 6, 7, 2, 5, 4$ get colours $1, 1, 2, 2, 1, 3, 4, 5$.

(b) Vertices $4, 5, 2, 7, 6, 3, 1, 8$ get colours $1, 2, 3, 3, 4, 2, 4, 1$.

5.15 Colour the vertices with colours $c_1, \ldots, c_4$. Then order vertices so that all coloured c_1 come first, then those coloured c_2, etc.

5.16 All are class 1.

5.17 Colour edges of hamiltonian cycle using 2 colours. Remaining edges get a third colour.

5.18 (a) Use Theorem 3.1. (b) Any matching can have at most k edges so $\chi'(G) \geq \frac{1}{k}(k + \frac{1}{2})r > r$.

5.19 (b) Induction step. Remove a pendant vertex v: $T - \{v\}$ can be coloured in $\lambda(\lambda - 1)^{k-2}$ ways, and then v can be coloured in $(\lambda - 1)$ ways.

(c) In a colouring of G', x and y may get same colour, so subtract $f_\lambda(G'')$. For deduction, use induction on the number of edges.

(d) Recurrence $a_n = \lambda(\lambda - 1)^{n-1} - a_{n-1}$ gives $a_n - (\lambda - 2)a_{n-1} - (\lambda - 1)a_{n-2} = 0$. Auxiliary equation is $(x+1)(x - \lambda + 1) = 0$.

Chapter 6

6.1 $|A \cup B \cup C \cup D| = |A| + |B| + |C| + |D| - |A \cap B| - |A \cap C| - |A \cap D|$
$\qquad\qquad\qquad\quad - |B \cap C| - |B \cap D| - |C \cap D| + |A \cap B \cap C|$
$\qquad\qquad\qquad\quad + |A \cap B \cap D| + |A \cap C \cap D|$
$\qquad\qquad\qquad\quad + |B \cap C \cap D| - |A \cap B \cap C \cap D|.$

6.2 $100 = 67 + x - 44$ gives $x = 77$.

6.3 $100 = 70 + 49 + 49 - 20 - 25 - 35 + x$ gives $x = 12$.

6.4 $1000 - 142 - 90 - 76 + 12 + 10 + 6 - 0 = 720$.

6.5 (a) $|S_1| > 75, |S_1 \cap S_2| > 150 - 100 = 50, |S_1 \cap S_2 \cap S_3| > 75 + 50 - 100 = 25$,

   (b) $|S_1| = m(n-1), |S_1 \cap S_2| > 2m(n-1) - mn = m(n-2)$, etc.

6.6 Imitate Example 6.7: $\binom{102}{3} - \binom{51}{2} - \binom{61}{2} - \binom{71}{2} + \binom{10}{2} + \binom{20}{2} + \binom{30}{2} = 231$.

6.7 Take P_1: pattern 12 occurs,..., P_4: pattern 78 occurs. $N(i) = 7!$, $N(i,j) = 6!$, etc. Obtain $8! - 4.7! + 6.6! - 4.5! + 4! = 24\,024$.

6.8 Take P_i : $2i$ is in position $2i$ $(1 \leq i \leq 4)$. Get $24\,024$ again.

6.9 $|S| = 8 \times 7 \times 6 = 336$. Get $336 - 3(7 \times 6) + 3(6) - 1 = 227$.

6.10 Let S = set of all permutations; then $|S| = \frac{10!}{2^5}$. Take P_i: the two is are adjacent. Then $N(1) = \frac{9!}{2^4}, N(1,2) = \frac{8!}{2^3}$, etc. Get $\frac{10!}{2^5} - \frac{5 \cdot 9!}{2^4} + 10 \cdot \frac{8!}{2^3} - 10 \cdot \frac{7!}{2^2} + 5 \cdot \frac{6!}{2} - 5! = 39\,480$.

6.11 $\phi(n) = n - \sum \frac{n}{p_i} + \sum \frac{n}{p_i p_j} - \cdots = n \Pi_{p|n}(1 - \frac{1}{p})$.
$\phi(100) = 40$, $\phi(200) = 80$.

6.12 $f_\lambda(\mathcal{G}) = \lambda^n - \sum N(i) + \sum N(i,j) - \cdots$. $N(i) = \lambda^{n-1}$ (both ends of edge e_i get same colour). Also $N(i,j) = \lambda^{n-2}$. (There are two cases: e_i, e_j may or may not have a common vertex - get λ^{n-2} in both cases.) Also, $N(i,j,k) = \lambda^{n-2}$ if e_i, e_j, e_k form a 3-cycle, and is λ^{n-3} otherwise.

6.13 $S = \{(x_1, \ldots, x_{12}) : 1 \leq x_i \leq 6\}$. P_i : no x_i has the value i. Number of throws in which all numbers appear $= |S| - \sum N(i) + \sum N(i,j) - \cdots = 6^{12} - \binom{6}{1}5^{12} + \binom{6}{2}4^{12} - \binom{6}{3}3^{12} + \binom{6}{2}2^{12} - \binom{6}{1} = 953\,029\,440$. To get probability, divide by 6^{12}.

6.14 If S = set of all partitions into 4 parts, $|S| = S(10,4)$. Let P_i : $\{i\}$ is a singleton set. So $N(i) = S(9,3), N(i,j) = S(8,2)$, etc. Answer is $S(10,4) - \binom{10}{1}S(9,3) + \binom{10}{2}S(8,2) - \binom{10}{3}S(7,1) = 9450$. *Alternatively*, want number of partitions of type $2^3 4^1$ or $2^2 3^2$, i.e. $\frac{10!}{2^3 3! 4!} + \frac{10!}{2^2 2!(3!)^2 2!}$.

Chapter 7

7.1

$$
\begin{array}{ccc}
2 & 3 & 1 \\
3 & 1 & 2 \\
1 & 2 & 3
\end{array}
\quad\text{and}\quad
\begin{array}{ccc}
3 & 1 & 2 \\
2 & 3 & 1 \\
1 & 2 & 3.
\end{array}
$$

7.2 Round 1: 1 v 2, 3 v 4, 5 v 6. Round 2: 1 v 3, 2 v 5, 4 v 6, etc.

7.3 $\frac{1}{n}\sum_{i=1}^{n^2} i = \frac{1}{2}n(n^2+1)$.

7.6 $a_{ij} = a_{iJ} \Rightarrow 2i+j-2 \equiv 2i+J-2 \Rightarrow j \equiv J \Rightarrow j = J$.
$a_{ij} = a_{Ij} \Rightarrow 2i+j-2 \equiv 2I+j-2 \Rightarrow 2i \equiv 2I \pmod{n} \Rightarrow i \equiv I$ since n is odd. So A is a latin square. Similarly for B.
Orthogonality: $a_{ij} = a_{IJ}$ and $b_{ij} = b_{IJ} \Rightarrow 2i+j \equiv 2I+J$ and $3i+j \equiv 3I+J \Rightarrow i \equiv I \Rightarrow j \equiv J$.
Broken diagonal of A starting at a_{1j} consists of $j, j+1+2, j+2+4, \dots$, i.e. $j, j+3, j+6, \dots$. But $j+3u \equiv j+3v \Rightarrow 3u \equiv 3v \Rightarrow u \equiv v$ since $3 \nmid n$. So all entries in diagonal are different.

7.7 (b) M_3 is.

   (c) Suppose $a_{ij} = a_{IJ}$ and $a_{ij}^T = a_{IJ}^T$. Then $2i+j \equiv 2I+J$ and $2j+i \equiv 2J+I$. Subtracting gives $i-j \equiv I-J$ whence $J = I-i+j$. Thus $2i+j \equiv 2I+I-i+j$, i.e. $3i \equiv 3I$, so $i \equiv I, j \equiv J$.

7.8 (a) 1 appears an even number of times off the diagonal in symmetrically placed positions. So it appears an odd number of times altogether, so n is odd.

   (b) $a_{ij} = a_{iJ} \Rightarrow j(m+1) \equiv J(m+1) \Rightarrow 2(m+1)j \equiv 2(m+1)J \Rightarrow j \equiv J \pmod{2m+1}$. Similarly for columns. Since $2(m+1) \equiv 1 \pmod{n}$, ith diagonal entry is $2(m+1)i = i$. For $m = 2, A$ is

$$
\begin{array}{ccccc}
1 & 4 & 2 & 5 & 3 \\
4 & 2 & 5 & 3 & 1 \\
2 & 5 & 3 & 1 & 4 \\
5 & 3 & 1 & 4 & 2 \\
3 & 1 & 4 & 2 & 5.
\end{array}
$$

7.9 Orthogonality. Suppose $\lambda_i \lambda_k + \lambda_j = \lambda_I \lambda_k + \lambda_J$ and $\lambda_i \lambda_h + \lambda_j = \lambda_I \lambda_h + \lambda_J$. Subtracting gives $\lambda_i(\lambda_k - \lambda_h) = \lambda_I(\lambda_k - \lambda_h)$, i.e. $\lambda_i = \lambda_I$ since $\lambda_h \neq \lambda_k$, so $i = I$. Thus $\lambda_j = \lambda_J$, i.e. $j = J$.

7.10 The rows are the columns of (7.3) in a different order.

7.12 For example $3, 4, 2, 1$.

7.13 $\{2, 5, 6\}, \{2, 6\}, \{2, 5\}, \{5, 6\}$ have only 3 elements in their union.

7.14 Ignore suits; there are 4 cards of each value. Any k columns contain $4k$ cards which must have at least $\frac{4k}{4} = k$ different values among them. Apply Hall's theorem.

7.15 Any k A_i contain kn elements in their union. These kn elements must be distributed among at least $\frac{kn}{n} = k$ sets B_i, so the union of any k S_i contains at least k elements.

7.16 Take the 5×5 array N with $1, \dots, 25$ in natural order. The 5 rows and 5 columns give 10 blocks. Corresponding to A_1, get $\{5, 9, 13, 17, 21\}$, $\{1, 10, 14, 18, 22\}, \{2, 6, 15, 19, 23\}, \{3, 7, 11, 20, 24\}, \{4, 8, 12, 16, 25\}$. Similarly for $A_2, \dots, A_5$.

7.17 Take $A_i = \{(i, j) : m_{ij} = 1\}$ and $B_i = \{(j, i) : m_{ji} = 1\}$ where $M = (m_{ij})$. Apply Exercise 7.15 to obtain a permutation matrix P_1. (Alternatively apply Theorem 7.6 or Theorem 5.13.) Then repeat argument applied to $M - P_1$, to get another permutation matrix P_2. Then consider $M - P_1 - P_2$, etc.

Chapter 8

8.1 First round has games $\infty \,\text{v}\, 1, 9 \,\text{v}\, 2, 8 \,\text{v}\, 3, 7 \,\text{v}\, 4, 6 \,\text{v}\, 5$. For 9 teams omit games involving ∞.

8.2

$x_1 \,\text{v}\, y_2,$	$x_2 \,\text{v}\, y_3,$	$x_3 \,\text{v}\, y_4,$	$x_4 \,\text{v}\, y_5,$	$x_5 \,\text{v}\, y_1$
$x_1 \,\text{v}\, y_3,$	$x_2 \,\text{v}\, y_4,$	$x_3 \,\text{v}\, y_5,$	$x_4 \,\text{v}\, y_1,$	$x_5 \,\text{v}\, y_2$
$x_1 \,\text{v}\, y_4,$	$x_2 \,\text{v}\, y_5,$	$x_3 \,\text{v}\, y_1,$	$x_4 \,\text{v}\, y_2,$	$x_5 \,\text{v}\, y_3$
$x_1 \,\text{v}\, y_5,$	$x_2 \,\text{v}\, y_1,$	$x_3 \,\text{v}\, y_2,$	$x_4 \,\text{v}\, y_3,$	$x_5 \,\text{v}\, y_4$
$x_1 \,\text{v}\, y_1,$	$x_2 \,\text{v}\, x_5,$	$x_3 \,\text{v}\, x_4,$	$y_2 \,\text{v}\, y_5,$	$y_3 \,\text{v}\, y_4$
$x_2 \,\text{v}\, y_2,$	$x_3 \,\text{v}\, x_1,$	$x_4 \,\text{v}\, x_5,$	$y_3 \,\text{v}\, y_1,$	$y_4 \,\text{v}\, y_5$
$x_3 \,\text{v}\, y_3,$	$x_4 \,\text{v}\, x_2,$	$x_5 \,\text{v}\, x_1,$	$y_4 \,\text{v}\, y_2,$	$y_5 \,\text{v}\, y_1$
$x_4 \,\text{v}\, y_4,$	$x_5 \,\text{v}\, x_3,$	$x_1 \,\text{v}\, x_2,$	$y_5 \,\text{v}\, y_3,$	$y_1 \,\text{v}\, y_2$
$x_5 \,\text{v}\, y_5,$	$x_1 \,\text{v}\, x_4,$	$x_2 \,\text{v}\, x_3,$	$y_1 \,\text{v}\, y_4,$	$y_2 \,\text{v}\, y_3$

8.3 Yes, by Theorem 7.5.

8.4 (a) The spokes, (b) none, (c) edges $12, 34, 56$ in Figure 4.1(a), (d) "vertical edges".

8.5 In Figure 4.1(a) take 1236451 and 1652431. Each gives two 1-factors.

8.6 All.

8.7 Hamiltonian cycle gives two 1-factors; remaining edges form another.

8.8 (a) $A_i \,\text{v}\, B_{i+k-1}$ in round k, where suffixes are reduced (mod 5) to lie in $\{1, \dots, 5\}$.

(b) Use the join of the first two MOLS of Example 7.3 to get first round A_2 v B_3 on court 1, A_3 v B_4 on court 2, A_4 v B_5 on court 3, A_5 v B_1 on court 4, A_1 v B_2 on court 5, etc.

8.9 Follow the method of Section 8.3, using the first 3 MOLS A_1, A_2, A_3 of Example 7.3. Then use A_4 to determine the courts. Get following schedule where rows give rounds and columns correspond to courts.

$B_5 G_1$ v $b_1 g_1$ $B_1 G_4$ v $b_3 g_5$ $B_2 G_2$ v $b_5 g_4$ $B_3 G_5$ v $b_2 g_3$ $B_4 G_3$ v $b_4 g_2$
$B_4 G_4$ v $b_5 g_3$ $B_5 G_2$ v $b_2 g_2$ $B_1 G_5$ v $b_4 g_1$ $B_2 G_3$ v $b_1 g_5$ $B_3 G_1$ v $b_3 g_4$
$B_3 G_2$ v $b_4 g_5$ $B_4 G_5$ v $b_1 g_4$ $B_5 G_3$ v $b_3 g_3$ $B_1 G_1$ v $b_5 g_2$ $B_2 G_4$ v $b_2 g_1$
$B_2 G_5$ v $b_3 g_2$ $B_3 G_3$ v $b_5 g_1$ $B_4 G_1$ v $b_2 g_5$ $B_5 G_4$ v $b_4 g_4$ $B_1 G_2$ v $b_1 g_3$
$B_1 G_3$ v $b_2 g_4$ $B_2 G_1$ v $b_4 g_3$ $B_3 G_4$ v $b_1 g_2$ $B_4 G_2$ v $b_3 g_1$ $B_5 G_5$ v $b_5 g_5$

8.10 (a) Only sequences with no breaks are $HAHA\ldots H$ and $AHAH\ldots A$. But no two teams can have the same venue sequence (otherwise they couldn't play each other).

(b) ∞ and 0 have no breaks. Each other has a break next to its game against ∞.

8.11 Draw a bipartite graph with n black vertices labelled by the first round pairs and n white vertices labelled by the kth round pairs. Join a black to a white when the labelling pairs are not disjoint. By Theorem 7.6 with $m = 2$ there is a perfect matching, in which each edge represents a team. Play these teams at home in kth round.

Alternatively, apply Exercise 7.15 to the pairs of the first and kth rounds: choose a common SDR and play these teams at home in round k.

Chapter 9

9.1 (a) b not an integer. (b) $b = 14 < v$.

9.2 Complement of a $(13, 4, 1)$ design.

9.3 (a) $vr(k-1)\lambda - r(k-1)\lambda = \lambda(v-1)r(k-1) = r^2(k-1)^2$.

(b) $(k-1)\lambda + (v-k)\lambda = \lambda(v-1) + r(k-1)$.

9.4 Use $v\lambda = k^2 - k + \lambda$.

9.5 $(k-\lambda)^{v-1}$ must be a square, so, dividing by $(k-\lambda)^{v-2}$, which is a square, $k - \lambda$ must be a square. For final parts must check that the two designs would be symmetric.

9.6 (a) $(n^2 + n, n^2, n^2 - 1, n^2 - n, n^2 - n - 1)$,

(b) $(4m - 1, 4m - 1, 2m, 2m, m)$.

9.7 Use H_4 and Theorem 9.15.

9.8 Use Theorems 9.18 and 9.16.

9.9 In each case the differences are $\pm(d_i - d_j)$.

9.10 – 9.14 Check differences.

9.15 Given i, there are 10 associates in same row or column, and 5 associates given by L. So each block has 15 elements. Given i and j in the same row of N, they are in $4\,B_k$ where k is in the same row as i and j, and in B_h and B_ℓ, where $h(\ell)$ is the element of N in same row as $i(j)$ such that $h(\ell)$ and $j(i)$ have equal corresponding entries in L. Similarly for other cases.

9.16 (a) $B = 2A - J$ has 1 where A has 1 and -1 where A has 0.

(b)
$$\begin{aligned}
B^T B &= (2A^T - J)(2A - J) = 4A^T A - 2A^T J - 2JA + J^2 \\
&= 4(k - \lambda)I + 4\lambda J - 2(JA)^T - 2JA + J^2 \\
&= 4(k - \lambda)I + (4\lambda - 4k + v)J = vI \Leftrightarrow 4(k - \lambda) = v.
\end{aligned}$$

(c) $v = 36, k = 15, \lambda = 6$ satisfy $4(k - 1) = v$.

9.17 Orders $4, 8, 16, 32$ by doubling; $12, 20, 24, 44, 48$ by Theorem 9.18; 36 by Exercise 9.16; 40 by doubling 20; 28 by extending Theorem 9.18 to prime powers, using the finite field of 27 elements.

9.18 (9.9) becomes $S \geq \binom{N}{2}d$; (9.10) unchanged.

9.19 If $d(\mathbf{x}, \mathbf{y}) = e$ then $\mathbf{x} + \mathbf{y}$ has weight e; so min. weight $\leq$ min. distance. Conversely, if $w(\mathbf{x}) = w$ then $d(\mathbf{x}, \mathbf{0}) = w$; so min distance $\leq$ min weight.

9.20 Can assume $\mathbf{0}$ is in code. All sequences of weight 1 get corrected to $\mathbf{0}$. There are no codewords of weight 2 (by perfectness) so all sequences of weight 2 must get corrected to a codeword of weight 3. Use correspondence between subsets A of $\{1, \ldots, n\}$ and binary sequences $\mathbf{x}_A = x_1 \ldots x_n$ where $x_i = 1 \Leftrightarrow i \in A$. The 3-element subsets corresponding to codewords of weight 3 form an $STS(n)$: if $|B| = 2, B \subseteq \{1, \ldots, n\}$, B is in C where $\mathbf{x}_c$ corrects $\mathbf{x}_B$.

9.21 (a) Count in two different ways the pairs (A, P) where P is a pair of elements of B and A is a block other than B containing P.

(b) Count in two different ways the number of pairs (A, y) where y is an element of B and A is a block other than B containing y.

(d) Expanding $\sum_i (i - m)^2 x_i \geq 0$ gives $\sum i^2 x_i - 2m \sum i x_i + m^2 \sum x_i \geq 0$. Write i^2 as $2\binom{i}{2} + i$; then (a) and (b) give $k(k-1)(\lambda-1) + m^2(b-1) \geq (2m - 1)k(r - 1)$. Replacing $m(b - 1)$ by $k(r - 1)$ gives (d).

Next, in (d) replace $(b-1)k$ by $vr - k$ to get $(vr - k)(k-1)\lambda + vr(r - k) + k^2 - rk \geq k^2(r - 1)^2$. Now use Exercise 9.3(a) and take $r - k$ out as a common factor to get $(r - k)(r - 2rk + (k-1)\lambda + vr) \geq 0$. Finally use Exercise 9.3(b) to deduce that $(r - k)(v - k)(r - \lambda) \geq 0$, whence $r \geq k$, i.e. $b \geq v$.

Further Reading

The main ideas of the first two chapters are covered by most of the standard textbooks on combinatorics, e.g. Brualdi [5], Cameron [6] and the remarkably extensive book by Graham, Knuth and Patashnik [11]. Chapters 3 to 5 cover some of the basic ideas in graph theory; further details can be found in the many textbooks now available, of which we particularly note those of Diestel [9], Wilson [20], Wilson and Watkins [21], and the English translation [15] of the ever-fresh classic of König [14]. The reader who is interested in the history of graph theory is directed to the recent reissue of the book by Biggs, Lloyd and Wilson [4].

The inclusion-exclusion principle has excellent expositions in [5] and in van Lint and Wilson [18]. The proof of Cayley's theorem on labelled trees, using inclusion-exclusion, is due to J.W. Moon [19], and there is a very nice article by J. Dutka on the ménage problem in [10].

Latin squares are covered in great detail by Dénes and Keedwell [7], [8], and are the central theme of a recent book by Laywine and Mullen [16]. There is also material on latin squares in [1], [5], [6] and [18]. Connections between latin squares and tournaments are described in [1], where the topics of Chapters 7–9 are dealt with more fully. [1] also deals with block designs, as do [16] and [18], and there are good accounts of coding theory in the books by Baylis [3], Hill [12] and Van Lint [17].

Another important source of information is the web. Search for Fibonacci numbers or derangements or latin squares, and you will be lead to many fascinating and informative sites. Also freely available on the web is a short course on designs [2].

Bibliography

[1] I. Anderson, *Combinatorial designs and tournaments*, Oxford University Press, 1977.

[2] I. Anderson and I. Honkala, *A short course in combinatorial designs*, http://www.utu.fi/~ honkala/designs.ps

[3] J. Baylis, *Error-correcting codes: a mathematical introduction*, Chapman and Hall, 1998.

[4] N.L. Biggs, E.K. Lloyd and R.J. Wilson, *Graph theory 1736-1936*, Oxford University Press, 1998.

[5] R. Brualdi, *Introductory combinatorics*, 3rd edition, Prentice-Hall, 1999.

[6] P.J. Cameron, *Combinatorics: topics, techniques, algorithms*, Cambridge University Press, 1994.

[7] J. Dénes and A.D. Keedwell, *Latin squares and their applications*, English Universities Press, 1974.

[8] J. Dénes and A.D. Keedwell, *Latin squares: new developments in the theory and applications*, North Holland, 1991.

[9] R. Diestel, *Graph theory*, 2nd edition, Springer, 2000.

[10] J. Dutka, On the Problème des Ménages, *Mathematical Intelligencer* **8** (1986), 18–25.

[11] R.L. Graham, D.E. Knuth and O. Patashnik, *Concrete mathematics*, 2nd edition, Addison-Wesley, 1994.

[12] R. Hill, *A first course in coding theory*, Oxford University Press, 1986.

[13] G.A. Jones and J.M. Jones, *Elementary number theory*, Springer-Verlag, 1998.

[14] D. König, *Theorie der endlichen und unendlichen Graphen*, Akad. Verlag.,
      Leipzig, 1936.

[15] D. König, *Theory of finite and infinite graphs* (translated by R. McCoart),
      Birkhauser, 1990.

[16] C.F. Laywine and G.L. Mullen, *Discrete mathematics using latin squares*,
      Wiley, 1998.

[17] J.H. van Lint, *Introduction to coding theory*, 2nd edition, Springer-Verlag,
      1992.

[18] J.H. van Lint and R.M. Wilson, *A course in combinatorics*, Cambridge
      University Press, 1992.

[19] J.W. Moon, Another proof of Cayley's formula for counting trees, *Amer.
      Math. Monthly* **70** (1963), 846–847.

[20] R.J. Wilson, *Introduction to graph theory*, 4th edition, Longman, 1996.

[21] R.J. Wilson and J.J. Watkins, *Graphs, an introductory approach*, Wiley,
      1990.

Index

Lightning Source UK Ltd.
Milton Keynes UK
UKOW06f1552050117
291404UK00009B/350/P